MATHEMATICS ON THE INTERNET

A STUDENT'S GUIDE
1997 - 1998

LINK-SYSTEMS INTERNATIONAL
ANDREW T. STULL

PRENTICE HALL Upper Saddle River, NJ 07458

Editorial Director: *Tim Bozik*
Editor in Chief Science: *Paul Corey*
Associate Editor in Chief/Development: *Carol Trueheart*
Senior Editor: *Sally Denlow*
A.V.P. Production & Manufacturing: *Dave Riccardi*
Special Projects Manager: *Barbara A. Murray*
Manufacturing Manager: *Trudy Pisciotti*
Manufacturing Buyer: *Ben Smith*
Development Editor: *Shana Ederer*
Supplement Cover Manager/Designer/Illustrator: *Paul Gourhan*
Copy Editor: *Mindy DePalma*

© 1997 by **PRENTICE-HALL, INC.**
Simon & Schuster/A Viacom Company
Upper Saddle River, NJ 07458

All rights reserved. No part of this book may be reproduced or transmitted in any form or by any means, electronic or mechanical, including photocopying, recording, or any information storage and retrieval system, without permission in writing from the Publisher.

TRADEMARK INFORMATION:
America Online is a registered trademark of Quantum Computer Services Incorporated; CompuServe is a trademark of CompuServe, Incorporated; Microsoft Windows is a trademark of Microsoft Corporation; NCSA Mosaic is a trademark of the National Center for Supercomputing Applications; Netscape is a registered trademark of Netscape Communications Corporation; Java is a registered trademark of Sun Microsystems

The authors and publisher of this manual have used their best efforts in preparing this book. The authors and publisher make no warranty of any kind, expressed or implied, with regard to these programs or the documentation contained in this book. The author and publisher shall not be liable in any event for incidental or consequential damages in connection with, or arising out of, the furnishing, performance, or use of the programs described in this book.

Printed in the United States of America

10 9 8 7 6 5 4 3 2

ISBN 0-13-889833-2

Prentice-Hall International (UK) Limited, *London*
Prentice-Hall of Australia Pty. Limited, *Sydney*
Prentice-Hall Canada, Inc., *Toronto*
Prentice-Hall Hispanoamericana, S.A., *Mexico*
Prentice-Hall of India Private Limited, *New Delhi*
Prentice-Hall of Japan, Inc., *Tokyo*
Simon & Schuster Asia Pte. Ltd., *Singapore*
Editora Prentice-Hall do Brasil, Ltda., *Rio de Janeiro*

Contents

Preface	Change!...v	
Introduction	O' Brave New World: Brief Internet History........1	
Chapter 1	What's Under the Hood? The Basics..............5	
1.1	Gasoline and motor oil	- Connecting ... 5
1.2	What your mechanic never told you	- Organizing.... 10
Activity:	The Starting Line 13	
Chapter 2	Hit the Road Jack: Navigating The Net............15	
2.1	Cruising the net	- Browsing..... 15
2.2	Understanding the traffic signs	- Navigating.... 17
2.3	Asking for directions	- Searching..... 21
Activity:	The Big Road Trip...................25	
Chapter 3	Doing Your Own Tune-up: Getting Your Message Out. 27	
3.1	Fuzzy dice and seat covers	- Customizing .. 27
3.2	Horns, radios, and turn signals	- Communicating 32
3.3	Turbo chargers	- Extending 40
Activity:	A Talk-about Tour 43	
Chapter 4	Hot Rodding: Advanced Techniques 45	
4.1	There's no place like home	- Designing 45
4.2	Open your own shop	- Serving 56
Activity:	Make Yourself at Home 60	
Appendix		
I	It's News to Me	- Mathematics Newsgroups 61
II	Stepping Out	- Homepage Template.... 62
III	Useful URLs	- Mathematics Web Sites . 65
Glossary	It's all Greek to Me........................67	

Preface

Change! I wrote the first edition of this manual at 7:22 p.m. on July 23, 1995. It is now 2:04 p.m. on January 31, 1997 and a great deal about the World Wide Web and the world at large has changed. Dealing with change is a basic requirement for surviving in our modern world, and anticipating change may even make you rich. Our world is louder, faster, and more complex than the ones experienced by earlier generations. In terms of information transfer, we might be described as a techno-generation; our parents, as a paper-generation. In the past year and a half, the World Wide Web changed at an extraordinary rate, and it will probably continue to do so; our online future is likely to be chaotic but exciting. Prepare to revel in the difference that tomorrow will bring.

Reading this manual won't teach you all there is to know about the World Wide Web, but it will help you to teach yourself. In the future you will need to find information for yourself rather than relying solely on others, who may bear outdated knowledge. If you are successful, your skills in "cruising" the Internet will allow you to deal with perpetual change. By the end of this manual, you should be comfortable and resourceful in navigating the complexity of the Internet, from its back eddies to its thriving thoroughfares.

This manual has four chapters. In the Introduction, O' Brave New World, I will describe the origin of and the innovations behind the Internet. In Chapter 1, What's Under the Hood, we will explain the use of a Web browser and describe how you can obtain a connection to the Internet. The boxed and end-of-chapter exercises will give you practice in using your browser, as well as expose you to some of the wonderful places on the Internet.

In Chapter Two, Hit the Road Jack, you'll learn more about how to use your browser in order to cruise the endless byways of the Internet. Also, you will be introduced to resources and strategies for information searching. The boxed and end-of-chapter exercises will reinforce your navigational skills and give you practice at searching for some of the great mathematics resources available to you on the Internet.

In Chapter Three, Doing Your Own Tune-Up, you will learn about how to fine tune and customize your Web browser to make it more responsive to your needs. Also, you will find yourself changing from an observer into an enthusiastic Internet participant. The boxed and end-of-chapter exercises will help you reach out and make contact with others on the Internet.

In Chapter Four, Hot Rodding, you will move into the fast lane of Web publishing. We'll discuss the ins and outs of Web design, HTML editing software, and the nuts and bolts of putting together your own Web server.

Finally, in the Appendices a glossary of gearhead terms and a template for making your cruise more scenic are provided.

Acknowledgments

I'd like to thank my wife Elizabeth for the encouragement, the faith, and the unbelievable time she has spent editing and advising me on both this and the original edition of the guide. She has probably spent as many hours reviewing and providing invaluable suggestions for it as I have spent writing it. I am continually inspired by her inexhaustible energy and amazed by her skill for teaching and love for learning. I am also grateful to the following people for providing many helpful comments and suggestions:

>   Susan Brawley, *University of Maine*
>   Ron Edwards, *University of Florida*
>   Patrick A. Thorpe, *Grand Valley State University*
>   Robin Tyser, *University of Wisconsin—LaCrosse*
>   Roberta Williams, *University of Nevada, Las Vegas*

Finally, I would like to thank both Sheri Snavely, my editor at Prentice Hall, and Paul Corey, Editor-in-Chief of Higher Education Science at Prentice Hall. They have allowed me to rewrite this guide and continue to listen to my rantings about the direction of education on the Internet. I am both pleased and amazed that they continue our relationship with such patience.

Introduction
O' Brave New World

This thing that we now call the Internet has been evolving ever since it was first developed over twenty-five years ago. Many people have compared the Internet to a living creature because of the way it grows and changes. You may find its history quite interesting. Also, the reasons for its creation and growth are helpful in understanding the nature, terminology, and culture of the people who have adopted the Internet as home.

In the late 50s and early 60s, scientists and engineers realized the importance of sharing information and communicating through their computers. Many different groups attempted to develop computer language that would enable computers to exchange information with one another, or *network*. Most were successful. But ironically, all of these networks used different languages—people on different networks still had difficulty communicating with one another. It was like the Tower of Babel all over again.

The Internet was born as the solution to this problem. The U.S. government paid for the development of a common network language, called a *protocol,* which was eventually shared freely. Over time, many formerly isolated networks from all over the world adopted this language. Thus, the best description of the Internet is that it is not a network, but a network of networks. However, the Internet is independent of governments and regulation—there is no Central Internet Agency. Change is spurred by the common needs of the people that use the Internet.

Admittedly, this type of network system isn't the most graceful—but it works. If you saw a diagram of this great big computer network, you might find it resembles a spider's web. On this web, information can travel between any two points along any one of many possible paths.

Originally, the chief purpose of the Internet was to provide a distribution system for scientific exchange and research. Gradually, however, the Internet also became a digital post office, enabling people to send mail and transfer computer files electronically. Although the Internet is still used extensively by scientists, the commercial sector is currently the most powerful force behind its growth.

Historically, as technology changed, the speed with which information could be transferred and the way we viewed information changed. In 1991, an important new user interface was developed at the University of Minnesota: the *Gopher. Gopher* is a visually-oriented search tool for the Internet that allows users to locate information found on other computers. Because of *Gopher,* and other, more sophisticated *graphical user interfaces* developed since 1991, it is now possible to search through vast stores of information on computers all over the world. Once the desired information is found, it can be easily downloaded to the user's computer. Amazing if you think about it! You could be on your computer in Toledo, Ohio, and view information from Hamburg, Germany; Mexico City; or Tokyo, Japan without even realizing it. Wham! And no airline tickets!

In 1992, researchers in Switzerland helped to create a new format for information exchange that led to the explosive growth of the World Wide Web (WWW). Information on the Web is posted as a "page" that may contain text, images, sounds, and even movies. The organization of a page is much like any printed page in a book. However, web pages make use of *hypermedia*. Hypermedia involves the use of words and images as links, or connecting points, between different texts, images, sounds, or movies on other computers throughout the world. *Hypertext* web pages contain links only to other text documents.

However, the introduction of the Web created a dilemma: It was a great place to go, but there was no easy way to get there (kind of like the moon in the 60s.) We still lacked a convenient software program that would allow users to access the Web easily. In 1993, a program called *Mosaic* was developed by the National Center for Supercomputing Applications (NCSA). It allowed the user to browse Web pages as well as use other Internet resources such as electronic mail (e-mail).

After this browser was released, the Web has began to grow faster than the speed of light. In 1991, around 700,000 people were using the Internet. After *Mosaic* came out, users increased to around 1.7 million. The release of another innovative browser, *Netscape Navigator,* took place when users were estimated at 3.2 million (July 1994). Since then, the growth hasn't slowed much—conservative estimates suggest that over 10 million people have access to the Internet; radical reports place the number at 100 million.

Today, the Internet is changing at staggering rates and becoming more readily available to the average person. Just listen to the computer jargon in the popular media. When was the last time you saw a movie, heard a radio commercial, or read a magazine without encountering something about the Internet? As I write this sentence my radio station is talking about B. B. King's new Web tour and virtual concert on the Internet.

Today, you have access to animation, video clips, audio files, and even virtual reality worlds. Imagine all the new ways we will be able to view tomorrow's digital world.

> For those of you who already have some Web experience, here are a couple of Web addresses discussing the history and growth of the Internet. Simply type the address into your Web browser exactly as it appears below. If you are relatively new to the Internet, you can refer to Chapter 1 to learn more about Web browsers and Internet addresses.

BBN Timeline

BBN includes an Internet history timeline. It places the important Internet events in context with other historical events and throws in plenty of social commentary to give you perspective.

Address: http://www.bbn.com/customer_connection/timeline.htm

Hobbes' Internet Timeline

Hobbes' site offers a great deal about the Internet, the people who use it, and online culture.

Address: http://info.isoc.org/guest/zakon/Internet/History/HIT.html

Netizens: On the History and Impact of Usenet and the Internet

This is a comprehensive collection of essays about the history, nature, and impact of the Internet.

Address: http://www.columbia.edu/~hauben/netbook/

If you are just beginning to learn about the Internet, you might want to visit these sites later on.

Chapter 1
What's Under The Hood?

Many of you reading this manual have a lot of experience with computers, while others have little or none. In the first section of this chapter, I will briefly describe the basic computer setup you'll need, how to use a modem, and choose an Internet Service Provider (ISP). Many of you may be lucky enough to have computers on your campus that are set up to allow Internet access. In case you don't, I'll list the minimum in terms of systems, connections, and services that you'll need for the Internet. There are a staggering number of computers, software, and connections that you can use to get onto the Internet.

In the second section of this chapter, I will explain some of the idiosyncrasies of the Internet and describe the general features of most Internet browser software. A popular Web browser, *Netscape Navigator,* is used to illustrate discussions. Another popular browser is *Internet Explorer* from *Microsoft.* Both are *free* to students. I don't advocate any particular browser; you will probably want to try various browsers and make up your own mind. Although our illustrations focus on *Netscape Navigator*, fear not; both browsers share many of the same features and once you've learned the basic techniques, it's easy to switch back and forth.

Section 1.1
Gasoline and motor oil

What is the difference between a Viper and a Geo? Okay, it might seem like a silly question, but give it some consideration. The main difference is in price (the Viper is much more expensive). But if we consider how well each of these cars meets our basic need for transportation, the two cars are very similar. The same goes for computers and networks. The simple no-frills stuff will save you money while taking you where you want to go; the high-gloss stuff will transform your cash into dash and make your Internet browsing a little more enjoyable.

To get started, you'll need a *computer,* a *modem,* an *Internet connection*, and *browser software*. The descriptions that follow will help you understand each component and its function as you set up your own Internet access.

The Computer
Be careful how you approach this issue if you ask someone for advice on which computer platform to buy. Many people have strong opinions about the differences between Macintosh and PC-compatible systems. The best advice that I can give to you is test them

both at a computer store. Choose the one that you can pay for and are most comfortable using. After all, it won't do you any good if you don't enjoy using it.

These are the *minimum* system configurations that you'll need.

Macintosh	PC-Compatible
68030	Intel 486
System 7.0	Windows 3.1
256 color monitor	VGA monitor
16 MB of RAM	16 MB of RAM
8 MB of free disk space for browser software	8 MB of free disk space for browser software

A new innovation is the Network Computer, or NC. An NC is a computer without all of the things that you would expect in a computer: word processing, drawing, graphing, and number crunching. Because these features may be helpful to your education, you should consider the purchase of an NC carefully. Similar products will allow you to connect your television directly to the Internet. *WebTV* is currently the most popular, but I suspect that you'll see many different brands in the near future. The advantage to such products is that they are much cheaper than a full blown computer and you don't have to be a computer genius to use them.

The Modem

So—you probably want to know why you need a modem if you already have a computer. A modem is a device that MOdulates and DEModulates—that is, it translates a computer signal into a telephone signal, and vice versa. Although computers and telephones were set up to speak different "languages," you can use a modem to translate between your computer and another computer across your telephone line. Modems come in different "sizes," so don't just go out and buy the cheapest one on sale. Definitely don't buy one from a garage sale unless you really know what you're doing. Because modem technology changes so quickly, older equipment may be useful only as a doorstop. The number one thing that you need to know about a modem is its speed of transmission. Modem speeds are referred to in units called baud (a bit is a basic unit of digital information and a baud is the speed of transmitting 1 bit in 1 second). At one time a modem speed of 2600 baud was considered adequate. However, the minimum speed requirements have been steadily increasing as users demand more information at faster rates. You should purchase a modem with a speed of at least 28,800 baud (28.8k baud). With a typical 28.8k baud modem you can expect that it will take a few seconds to transfer a typical Web page. However, keep in mind that manufacturers will continue to introduce newer and faster modems as pages become more complex and slower to load, and as users demand faster speeds.

You also need to make sure that your modem will work with your computer's operating system. Generally, this isn't a big deal, as all modems are basically the same and top manufacturers produce software for all of the major operating systems. Just remember to read the box to make sure the software you need is included. Unfortunately, explaining how to install the modem software would require more pages than my editor will allow. Not to worry. Included with the software is an installation manual and a phone number to a help desk. If you run into trouble, don't hesitate to try both. As for which brand of modem to purchase, I can only tell you to buy what you can afford. Your internet service provider or your campus computer administrator may recommend a particular brand of modem. Take this suggestion seriously, as the technicians within your ISP or campus are probably more familiar with the recommended modem and will be able to help you with ease if problems arise.

The word *modem* may also refer to a device that allows you to connect your computer or television to a service line. By the time this guide is published, you will undoubtedly hear of things called ISDN modems and cable modems. An ISDN modem is a classic misnomer because the ISDN signal is already understood by computers and isn't modulated and doesn't need to be demodulated. The cable modem refers to a box that connects between your cable TV line (not your telephone line) and your computer or television.

The Internet Connection

If you're lucky, your campus has already recognized the importance of the Internet as a teaching and learning tool. If the equipment is set up on-campus, then you may already have access to the Internet, or more specifically, to the Web. Otherwise, many resources exist to help you set up a connection from home.

Some campuses, although lacking a walk-in lab, have made arrangements for students to dial into the campus computer system and connect to the Internet with a modem. If this is the case, then check with one of the campus computer assistants or the campus computer hack. Advice from a hack may be extremely useful—he or she could be an excellent source of information.

Another option is to subscribe to a company such as *America Online, Compuserve, Prodigy, Microsoft Network,* or one of the many independent "mom and pop" companies currently offering monthly access to the Internet. It is a buyer's market and you should shop around. Test drive everything before you buy. This will save you a great deal of frustration. Here are a few things to consider when choosing an internet service provider.

> **Does the ISP have a local number for your area**? You need to call the provider each time you access the Internet. Paying a toll call every time you do so will cost you a ton of money if you use the Internet regularly.

Can their system handle a large number of simultaneous connections? Ask them how many users they can handle at one time and how many subscribers they have. Although they may have a reasonable price and a local number, it doesn't mean much if you can't get on to use it. If after you subscribe you find that you are never able to connect or that the only available access is late at night or early in the morning, then find a new ISP.

Do they offer SLIP/PPP connections? This is the type of connection that you'll need if you want to use a graphical browser like *Netscape Navigator* or *Internet Explorer*. Some ISP's only offer shell accounts. Shell accounts require you to type in each command as you would with DOS. It is somewhat like driving a horse and buggy when everyone else has an automobile.

Do they have a reasonable monthly subscription fee? Cheapest is not always best. The added features and the staffing support are important points to consider when choosing a service. Some internet service providers offer you unlimited monthly connect time at a flat fee and others offer you a per hour fee with additional hours being extra. You will need to guesstimate your expected usage and purchase accordingly. Ask if there is a fee to upgrade your service if you find that you need more time. If you have a roommate, then consider upgrading the service and splitting the cost. This may actually save you money.

Does your ISP include the Internet browser software in the price? You'll find that not all do. Most ISP's have an agreement with either *Netscape Navigator* or *Microsoft* to bundle their browser software. The provided software may also be partially configured to work on the ISP's system, so you'll be much farther along by using it and the technicians will be better able to help you with a problem.

Is the ISP a regional or local company? This may not be important to everyone, but some of you may go home during holidays and vacation. If the ISP covers a wider area, then you can still check your e-mail and cruise the Net when you are away from school.

Do they have a help line in case you need technical assistance to set up your connection? Call the help line before you subscribe and make sure you get a real person. Although you may be asked to leave your name and number, you should expect to get a return call within 24 hours minimum. If they don't return your call within this time period, then the service is probably understaffed or poorly managed.

Does the ISP offer both newsgroup and e-mail access in addition to a connection to the Web? This is usually standard but there are always exceptions; it is better to ask up front.

Does it cost you extra for additional e-mail addresses? If you have a roommate, then you may find that it is more affordable to split the cost of a subscription and pay for an additional e-mail account.

Will your ISP add newsgroups at your request? Most ISP's subscribe to a small fraction of the available newsgroups and you may find that they don't include some of the basic, science-oriented groups that your instructors may recommend. It shouldn't cost anything for the ISP to add these groups to their list.

Does the ISP offer you space for your own Web page? Often, one of the features offered in the basic package is the option of constructing and posting your own page. The ISP usually sets a memory usage limit that affects the total size of the page and its traffic flow (that is, the number of people viewing the site).

The most important thing to remember when using an ISP is to expect courteous and prompt service. If you don't like what you are paying for, then cancel and go somewhere else. There are plenty of competitors willing to offer you better service.

The Browser Software

A descriptive name for software such as *Netscape Navigator* or *Internet Explorer* is *browser,* as that is what most people do with it. It is used to browse or wander, sometimes aimlessly, through the Internet.

Many Web browsers are on the market today, and new ones frequently enter the race to capture your dollar.

All browsers have advantages and disadvantages. You should evaluate several and choose the one that you're most comfortable with. (However, when choosing a browser, remember that seeing over the dashboard is all that is really important. Don't get wrapped up in features that you'll never use.) Browsers are typically very cheap, if not simply free, for educational use. *Netscape Navigator, Internet Explorer*, and many of the other browsers are FREE for student use! So, don't be afraid to look at several. At the end of the chapter you will find several Web addresses that offer such software. Of course, if you purchase it at the store, you also get a user's manual, which you don't get with the free, educational-use copy.

Now, if you have all of these basic elements and they've been put together correctly, you should be ready to surf. If you still haven't put all of the pieces together, then jump to the activity (*The Starting Line*) at the end of the chapter for a step-by-step approach. The rest of this manual is devoted to explaining some problems you might encounter as you explore the many strange and wonderful places on the Internet. As you explore, you will become more comfortable and better able to utilize the Internet for both education and entertainment.

Section 1.2
What your mechanic never told you

The software that you'll use to access the Net is commonly called a *Client* or a *Web browser*. It functions according to an information exchange model called the *Client-Server model* (Figure 1). In this way, a *client* (your Web browser software) communicates with a *server* (a computer with Web server software) on the Internet to exchange information. When referring to the Web, the information that your browser receives from the server is called a *page*.

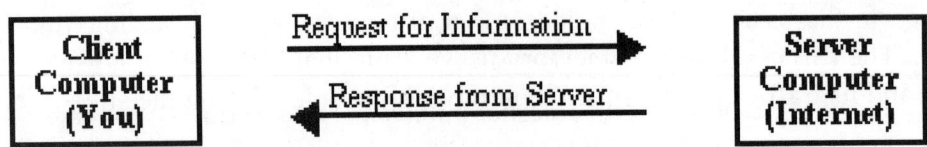

Figure 1.1 The Client requests information from the server. The requested information is displayed on the client computer.

So, what really appears on these Web pages? The best way to find out is to see for yourself. If you can, sit down in front of a computer, start your browser software, and connect to the Internet. Your browser is probably already set to start at a specific page. This start page is often referred to as a *homepage*. Web pages usually include both text and images. Some also use sounds and videos. A basic rule of the road is that anything that can be saved or recorded onto a computer can be distributed on the Internet through a Web page. The use of different information types is called multimedia.

When you choose a page it will be sent to your computer. After the requested information has been sent by the server, your computer will display it for you.

Soapbox Racers
Sometimes the Internet is not quite as responsive as you'd like. Here is an analogy that might help you understand the workings of the Internet a little better. Did you ever build a soapbox racer when you were a little kid? They are not as popular now but you can still find a derby if you look hard enough (Here's a Web address if you want to get started in soapbox racing: http://pages.prodigy.com/SOAPBOX/ - I'll explain how to use this later.).

Imagine this: You hear about a derby and get together with some friends to build a soapbox racer. Everyone brings something (a rope, a board, some nails, spare bike parts, etc.). With these scavenged parts, you begin to prepare your racer for the big soapbox derby. When the big day comes and you arrive at the race, you realize that the other racers were made from spare parts and scrap too. Although they all look somewhat alike, they are also still quite different.

Well, think about it. The Internet is like a soapbox derby. It is made up of hundreds of networks patched together. Every once in a while a wheel falls off during a race.

Traffic Jams and Construction Work

So, why is the Internet so slow sometimes? Think of the Internet as a big spider web. If you're the spider and you're trying to get to a fly stuck in the web, you usually have more than one path to get there. Some paths are more direct than others, but there are choices. Like a spider web, sometimes a small section of the Internet drops out of the "Web" and traffic has to be rerouted. This obviously causes increased traffic on the remaining strands, which in turn increases your waiting time.

Local or Long Haul

You need to also remember that the Internet mimics the real world. Distance is a factor in determining how long it takes to access a Web page. Generally, loading a page from a machine across town is usually much faster than loading a page from across the nation or across the world.

Time Zones and Tuna Sandwiches

The world works on different time zones and the Internet does too. Now, let's talk about lunch. Most people in the Western world take lunch around noon, and a many of them check their e-mail or browse the Web as well. So, you can usually expect Internet traffic to be slow during that time. Let's now consider the distance factor. There is a three hour time difference between the East and West Coast of North America, so the lunch rush lasts about four hours. Your location will determine if you are on the lead, middle, or tail end of the rush. Plan accordingly.

The Parking Lot

Surfing the Web can be like shopping during the holidays. You either arrive early or park a few miles away. Here's the connection: A transaction occurs between your computer and another when you load a Web page for viewing. You require a document (a Web page) from somewhere on the Internet (a server). Obviously, a slow connection to the Internet on the client side may cause delays. But consider what is happening on the server end of the transaction. Small, slow servers will take longer to serve Web pages than large, fast servers. Now think about the holiday rush: Although there is normally adequate parking, a holiday sale and a limited number of parking spaces can add hours to your shopping. It should not be difficult to see how large, fast servers can be rapidly overloaded if they are hosting a really interesting Web site.

When Microsoft released an updated version of *Internet Explorer,* over twenty state-of-the-art machines went down because of excessive demand. Basically, *Microsoft*'s parking lot wasn't big enough for the shopping rush.

Sometimes Sh*t Happens

You may notice that sometimes things on the Internet don't make any sense. It helps to remember that you're playing in a soapbox car. Here's an example: I live in southern California and one day I was attempting to retrieve something from the East Coast. I couldn't get through. I contacted the administrator of the server and they verified that the server and material were available. Yet I still couldn't get through. With a little research I discovered that everything east of Ohio and north of Delaware was blacked out for me. As it turned out, on that particular day, a major wheel on the Internet soapbox car had fallen off.

By this time, you probably understand enough about each of the basic components of an Internet-ready computer to make a start. Even if you're feeling intimidated, don't worry. You don't need to know everything there is to know about each component. And you can always learn more from the user's guides provided by the manufacturers. Just take it one step at a time and you'll piece everything together.

The following activity is a check list to help you progress through the steps for connecting to the Internet.

Activity: The Starting Line

A key to a car is of no use without a car, some gas, and a road to drive on. But remember, you really don't need a planned destination to have fun. This activity is intended to help get your computer on the road. Even if you don't have a working computer on the Internet, try to use a friend's or one on campus to view this material.

The Browser

Right now you should have a computer and a modem. Use the following Internet addresses to research the right browser for you.

Netscape
http://www.netscape.com
Microsoft
http://www.microsoft.com

The Internet Service Provider

Now, with your computer, modem, and browser, you're only one step away. Use the following Internet addresses to research the right ISP for you. You might want to consider the questions I outlined previously.

Choosing the Internet Service Provider Netscape
http://home.netscape.com/assist/isp_select/index.html
Internet Access Provider Guide
http://www.liii.com/~dhjordan/students/docs/welcome.htm
Choosing An Internet Provider
http://tcp.ca/Dec95/Commtalk_ToC.html

Research

It is always nice to have an independent opinion; therefore, read what the critics have to say. The following Internet addresses are for two of the largest publishers of computer related magazines. Between them, they print nearly 50 different popular periodicals about computers and the Internet. Search their databases for articles that will help you decide. You can read the articles online.

CMP Media Inc.'s (Publisher of Windows Magazine and others)
http://www.techweb.com/info/publications/publications.html
Ziff Davis (Publisher of PC Magazine, MacUser, and others)
http://www5.zdnet.com/findit/search.html

Chapter 2
Hit the Road Jack

By now, you should be prepared to see the world, or at least the World Wide Web. In this chapter, you'll develop a better understanding of the browser as a tool for navigating the Internet. The first section discusses basic navigation techniques; the second describes how to read Internet addresses; and the third introduces you to some of the information tools available on the Internet for finding your way around.

Section 2.1
Cruising the Net

Although the Internet may seem large and disorganized, finding your way around is no harder then finding your way to a friend's house. Information on the Internet has an address just as your friends do. Most browsers allow you to type in an address and thereby access information from, or "go to", a particular document.

Let's take a look at how to enter an address using the Web browser *Netscape Navigator*, shown in Figure 2.1. Start by finding the text entry box, which is located to the right of the words GoTo. (Sometimes the word *Location* is used instead.) If you have *Netscape Navigator*, type in the address and press the return button on your keyboard.

Figure 2.1. The tool bar for *Netscape Navigator* offers basic navigation features such as *Back, Forward,* and *Home.* You will find everything you'll need to perform simple navigation around the Internet.

Now take a look at the row of boxes directly above the text box. Each one contains an icon; together, they are known as the *tool bar*. Clicking on an individual box causes the computer to execute the command noted within the box.

How do each of these commands help you to navigate the Internet? Let's return to our analogy. Suppose you go to a friend's new home for a party but forget the house-warming present. What do you do? Drive home, of course. A Web browser will let you do something similar. In Figure 2.1, find the button labeled *Back*—it's at the far left. By clicking on this button you can return to the Web page you just visited. If you have gone to many Web pages, you can use it repeatedly to make your way back to your starting point.

> Here's the address of a site you might want to visit. When you type it into your browser, you'll arrive at NewsLinks, which is published by Simon & Schuster. It's not free but it is very reasonable and very good.
>
> http://www.ssnewslink.com/
> NewsLink is a subscription service that you can use to get news delivered about current issues and events in science or other topics. There are other news pages on the Web, many of which are free.

Now find the *Home* button. By selecting this button, you will immediately return to the homepage configured for your browser. When you first begin using your browser, it will be set to a page determined by the company that created it. Later on, you'll learn how to make any page you wish into a homepage, and even to create your own. But for now, remember that you can always go home.

Two other buttons common to most browsers are *Stop* and *Reload*. *Stop* is pretty easy to understand but *Reload* needs just a bit more explanation. As you explore more of the Internet, you'll realize that complete pages don't show up in your browser window instantly. Instead, different types of elements (pictures, icons, text, animations, etc.) appear over time as they are moved from the server to your browser. Occasionally, you'll notice that a page loads without some of these elements. This is often caused by an error in the transmission. Use the *Reload* button to request a new copy of the page.

You don't always have to know the address of a page to view it. The wonderful thing about the Web is that you can access (navigate) pages through the use of *hyperlinks*. You will notice that hyperlinks are often colored words (typically blue) on a Web page. Images may also be hyperlinks. Your mouse is used to select or click on the desired hyperlink. Some Web authors write their pages so that their hyperlinks are hidden from you. If you aren't sure where the hyperlinks are, just click on everything. You can't break it. Clicking on a hyperlink will take you to a new Web page just as typing in an address does.

> Here are a few more places you might find both fun and interesting.
>
> **The Map Machine from National Geographic**
>   http://www.nationalgeographic.com/ngs/maps/cartographic.html
> **The Why Files**
>   http://whyfiles.news.wisc.edu/index.html

> **American Museum of Natural History**
>   http://www.amnh.org/
> **MTV On-line**
>   http://www.mtv.com/
> **Big Book**
>   http://www.bigbook.com/
> Use this opportunity to experiment with each of the toolbar buttons on your browser. Remember that you can always use your *Back* button and *Home* button to get you back to the beginning.

The author of a Web page can connect one image to many different places. This type of hyperlink is sometimes called a *clickable map*. For example, think of a Web page with an image of a flower. The creator of such a document could place a hyperlink under each flower piece (petal, sepal, stamen, anther, etc.) so that your selection of different pieces would take you to different information. For a little experience in using this type of navigation, visit the sites in the following box.

> Here are two addresses that use clickable maps. They are also great places to find science information.
>
> **The Subway from UC Berkeley**
>   http://www.ucmp.berkeley.edu/subway/subway.html
> **The Fractal Explorer**
>   http://www.vis.colostate.edu/~user1209/fractals/index.html
>
> Not all browsers support clickable maps. However, most good Web authors will provide conventional hyperlinks to related sites in addition to the clickable map.

Some Web pages contain *forms*. A form is generally a request for information, which you may respond to by clicking with your mouse or typing an answer. Forms can be used to survey users, answer questions, or make requests. If you'd like to see a Web page that uses forms, you might want to visit the Gallery of Interactive Geometry. At this site, you can use forms to build a rainbow, play a special kind of pinball or explore one of the many other interesting activities.

   Gallery of Interactive Geometry
       http://www.geom.umn.edu/apps/

Section 2.2
Understanding the traffic signs

The Internet addresses that you've been using are also called *Uniform Resource Locators*, or *URL*s. Each URL has a couple of basic parts just like a residential address. Look at some of the URLs (now you know what they are called) that you've used already. Do you notice any similarities between them?

> Here is another typical URL (this will take you to a pretty cool site). The three basic components of a URL are listed below the address. Compare this one to the others you've seen so far.
>
> http://www.riddler.com/bridges/genericbr.html
>
> | | |
> |---|---|
> | **protocol** | http:// |
> | **server** | www.riddler.com |
> | **path** | /bridges/genericbr.html |
>
> You may notice that some addresses don't have a path element--this information is not always necessary. Be careful when you type a URL. Even one incorrect letter will prevent the browser from finding the desired site. Here is an additional resource if you'd like to learn more about Internet addresses:
>
> http://www.ncsa.uiuc.edu/demoweb/url-primer.html

Now pick up any recent magazine and leaf through the articles and advertisements. With little effort you should be able to recognize a few more URLs. Just as you might recognize a string of numbers to be a phone number (e.g., 555-1212), you should be able to spot URLs by their characteristic form and order. A URL may seem confusing at first glance, but think of it as a postal address squished together without any spaces.

The *protocol* of the URL indicates how the information is stored. In the exercise above, *HTTP* refers to the protocol, or language, that is used by all Web servers. The colon and slashes are used to separate it from the name of the server. They are not necessarily present in every type of URL.

> Up to this point, we have only been discussing browsers as an interpreter for documents that use hyperlinks. However, browsers also have the power to link to other, much older, formats of information. Try the following URLs to notice how the information differs.
>
> ftp://nic.merit.edu/documents/rfc/   (read file rfc0959.txt)
> gopher://gopher.micro.umn.edu:70/1
>
> HTTP, FTP, and Gopher are not the only protocols, but they are the three you're most likely to encounter. Have you ever encountered the following Internet protocols?
>
> telnet://
> wais://
> news:

The *path* describes the location of the Web page on the server. A *domain* is just a fancy name for a functional network group. The last part of the server name defines the domain to which the server belongs. The most common domain is indicated by the letters *edu*; all educational institutions are members of this domain group. Another important domain is *com*, which stands for *commercial* and includes Internet servers that belong to commercial companies. You are also likely to encounter domains that serve other groups; abbreviations for a few of these are shown below.

> Here are some URLs that may help expand your understanding of domains. Check them out. Can you determine what each domain group stands for?
>
> **http://www.microsoft.com**
> **http://www.nasa.gov**
> **http://www.navy.mil**
> **http://www.envirolink.org**
> **http://www.internic.net**

Always focus on reading URLs accurately. Suppose that you told a friend that your Mom lives at 123 Home Street, but the correct address is 123 Home Court. Your friend might have trouble finding the house. Likewise, a Web address must be accurate. There is a difference between http://www.mom.edu and http://www.mom.com--the first Mom wants you to get a good education and the second Mom wants a return on her investment.

In your travels, you will eventually jump to a server that is outside your country. Much, but not all, of the information on these servers is in English. URLs of foreign servers have an additional section at the end of the server name. It is a two letter code that indicates the country. Here are just a few examples of the many you might find.

    .au   Australia
    .ca   Canada
    .ch   Switzerland
    .nl   Netherlands
    .pe   Peru
    .uk   United Kingdom

> Here is yet another opportunity to encounter new places on the Web. It is a search tool that allows you to search for specific things. This group of tools is discussed later in the chapter, but you might want to play with it first. Enjoy!
>
> **Yahoo!**
>   http://www.yahoo.com

> Yahoo! is a directory of information that was originated at Stanford University. Remember you won't break anything. If you get lost, you can always shut down the program, have lunch, and try again later!

Return Visit

Now that you've found so many really interesting pages, how do you find them again a day or a week later without going through the same hassle? Your browser provides a way to mark an intriguing site and quickly return to it. *Netscape Navigator* calls this feature *Bookmarks* (see Figure 2.2). By using such a feature, you can compile a list of pages you frequently visit. When you wish to return to one of these pages, you simply select it from the list by clicking on it with your mouse. You will jump directly to the site.

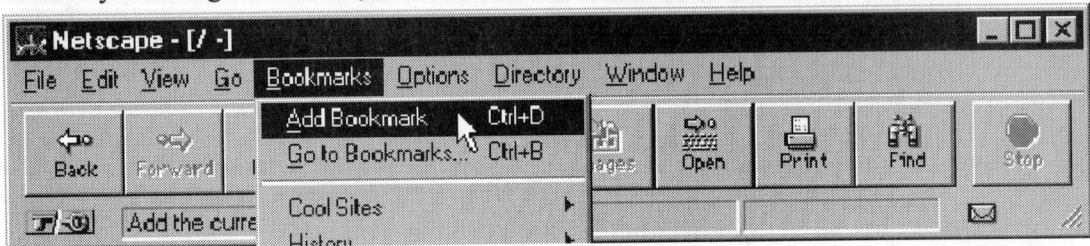

Figure 2.2. As pages are added to this list they will appear at the bottom of this same pull-down menu.

Here is a URL to get your collection started. Try this address: **http://www.hotwired.com**. It's a pretty interesting place: the online version of *Wired Magazine*. If you have Netscape, select *Add Bookmark* from the *Bookmark* menu (see Figure 2.2). Later on, if you use your mouse to select the *Bookmarks* menu, you'll see that *Hotwired* is just a jump away. You don't have to memorize the URL or haphazardly jump around until you find it again.

> Now that you've got the hang of it, give the following URLs a try.
>
> **ESPN Sports**
>   http://espnet.sportszone.com
> **Independent Underground Music Archive**
>   http://www.iuma.com
> **NBC (National Broadcasting Company)**
>   http://www.nbc.com
> **NFL (National Football League)**
>   http://www.nfl.com
> **Time Warner Communications (Life, People, Sports Illustrated, and more)**
>   http://www.pathfinder.com
>
> When you find something that you think is fun and interesting, save it or give the address to a friend.

As you begin to accumulate more sites on your list, you'll notice how cluttered and disorganized they can become. In addition to the ability to collect Web addresses, most browsers also allow you to organize them as you desire. Refer to Figure 2.2 and notice the menu option immediately below the *Add...* feature. This option, called *Go to Bookmarks*, allows you to organize your Web address list. When you select this option, a small window will open; in it you can move and manipulate your list of Web addresses.

History

If you've been working along with your manual and your browser at the same time, you've probably been to many places on the Web. Obviously, if you want to move back and forth between your selections, you are probably using the *Forward* and *Back* buttons on the browser toolbar. But what do you do if you want to go back to someplace you've visited fifteen jumps ago? Do you press the *Back* button fifteen times? Well, it works, but it is a little slow. You might want to try a new feature: It allows you to jump rapidly to any of the places you've visited along your path of travel.

In *Netscape Navigator*, select the *Go* menu. You'll see a list of all of the places you've visited on your journey. You need to select only the site you wish to jump to. With *Netscape Navigator*, you'll also find a similar feature called *History* under the *Windows* menu. By selecting this, you'll open a new window listing the names and the URLs for all of the places you've been to recently. Other browsers have similar features.

Because the list compiled under *Go* or *History* begin anew each time you start your browser, you should use *Bookmarks* when you find something interesting.

Section 2.3
Asking for directions

The Internet is an information jungle: It contains plenty of valuable information, but it's not always easy to find what you need. However, a number of easy-to-use tools are available for free. With practice, these tools will help you develop better information-finding skills.

There are two types of tools that you'll use most often to search for information. These are *Directories* and *Search Engines*.

Directories

Yahoo! is only one of many directories that are available on the Internet, but it is the best one available for general topics. I didn't name it and I don't know how it was named, but it is easy to remember. Here is the URL for *Yahoo!*:

      http://www.yahoo.com/

Yahoo! began as a simple listing of information by category—kind of like a card catalog. In the last two years, it has added the ability to search for specific information. At the top level of the directory, there are several very general categories, but as you move deeper into the directory you'll notice that the categories become more specific. To find information, you simply choose the most appropriate category at the top level and continue through each successive level until you find what you're looking for (or until you realize you're in the wrong place). Don't be afraid to experiment—it's easy to get lost but it's also easy to find your way home.

Suppose after visiting the *Fractal Explorer* mentioned earlier in this chapter, you decide to find out more about the mathematics related to fractals. Within *Yahoo!*, notice that one of the top level categories is Science. Science seems as if it would be the most appropriate place to find our information, so give it a try. After accessing the *Science* category, you'll notice that it gives a list of many different types of science. So, what is your next choice? My choice would be *Mathematics*, but *Chaos* is also a possibility (as you may have discovered during your visit to the *Fractal Explorer*, Chaos Theory is closely related to the study of fractals). Because *Yahoo*! cross references among the categories, you'll find that several related categories will lead you to your desired page.

Much of your success in finding information with this type of tool really centers around your preparation for the search. Often, it is possible to find information on a topic in a category that may at first seem unrelated to your topic of interest. Again, let's take the example of fractals. Although you may consider this a science-oriented topic, there are other avenues to consider. For instance, fractal images are often exceptionally beautiful and have developed into an art form. Categories related to art and computer generated graphics could also be searched.

Prepare yourself for a search before you jump into one. In the long run, it will save you both time and frustration. Don't be afraid to try some strange approaches in your search strategy. A good technique is to pull out your thesaurus and look up other names for the word. You might be able to find a more common form of the word. Think of everything associated with your question and give each of these subjects a try. You never know what might turn up a gold mine.

Search Engines

A more direct approach to finding information on the Web is to use a search engine, which is a program that runs a search while you wait for the results. Many search engines can be found on the Web. Some of Web search engines are commercial and may charge you a fee to run a search. Search engines are also available for other parts of the Internet: *Archie*, *Veronica*, and *Jughead* are examples of such search engines.

As mentioned earlier, *Yahoo*! also has a useful search engine. A search engine that I use frequently is called *Lycos (http://www.lycos.com)*. It's simple to operate but, as with any search tool, it takes practice and patience to master. Take the time now to connect to *Lycos*, and we'll take it for a test run. When you first see the opening page, you'll notice that it is very complex. But it's an excellent resource, and the instructions on the page will tell you almost everything you need to know. To search, enter a word into the white text entry box and press the submit button. *Lycos* will refer back to its database of information and return a page of hyperlinked resources containing the word you entered. When the query results come back to you, notice that they are hyperlinks to various sites on the Internet.

To see how a search engine works, use *chaos* as a topic for a search. Notice that you can set the number of responses that the engine will return to you. You may find that some of your results don't seem to apply to your topic. This is one of the downfalls of search engines. They are very fast, but they don't think—that is your job. For instance, a search using the term *chaos* may turn up a link to information about a popular game or a metaphysical site in addition to sites related to Chaos Theory. To perform an effective search, you will need to spend time before the search preparing a search strategy. When you do research using an automated tool like a search engine, you can expect many links to be unrelated to your topic of interest—but all in all, search engines are still very powerful tools.

One last word on search engines. These tools don't directly search the Internet. They actually search a database that is derived from the Internet. Here is how it works. Search engines use robots (automated programming tools) that search for and categorize information. This information is placed into a database. It is this database that you search when you use the search engine. Can you think of a potential problem with this system? Unfortunately, the quality of the database depends on the effectiveness of the robot that assembles the database. This is why you should not rely on just one search engine tool. Use several, because what one does not find, another might.

> Here's a way to test the skills you've developed from your accumulated Internet experience. Below are several words or phrases that are related to science and mathematics; try to find something related to them using *Yahoo!* and *Lycos*. Once you've exhausted these tools, try the other search tools listed at the bottom of this box. Think of it as a game—like the fox and hounds. Can you get the information before others in your class?
>
> | | |
> |---|---|
> | **Pascal's Triangle** | **Polynomials** |
> | **Prime numbers** | **Fibonacci numbers** |
> | **Convergent series** | **Graph Theory** |
> | **Mathematical modeling** | **Continuum Hypothesis** |

AltaVista Search Engine by Digital Electronics
  http://www.altavista.digital.com
BigFoot
  http://www.bigfoot.com
HotBot
  http://www.hotbot.com
WebCrawler
  http://www.webcrawler.com
Findspot
  http://www.findspot.com

Activity: The Big Road Trip

Here's what you've all been waiting for! You may want to find information on mathematics subjects covered in the popular press and media. The first list is a set of science and mathematics related topics that have been popular recently. See what you can find on these topics. Try to find the information in the least number of jumps possible.

> Fermat's Last Theorem
> Artificial Life
> Cryptography
> Knots and braids
> Complex Systems
> Mersenne Prime Search
> The Four Color Problem
> Symbolic computation
> Symbolic sculpture
> The Monty Hall problem

Learning through science doesn't always mean learning about discoveries in science. Sometimes it means learning about the scientists that make the discoveries. Hopefully, some of you will go on to be those great discoverers. A list of mathematicians from this century is found below. Some of these researchers will be very easy to track down, whereas others will not. I hope you'll find both their discoveries and their lives interesting.

Your assignment is to put together a brief history of each mathematician and their important discoveries. If you'd really like a challenge, try to draw a thread of connection between each mathematician, no matter how thin or obscure. (Hint: To help you understand the challenge, you might want to search out the term *Concept Map*).

> John von Neumann   Florence Nightingale David
> Alfred Tarski   Ruth Aaronsom Bari
> Vivienne Malone-Mayes   Alan Mathison Turing
> Kurt Friedrich Gödel   Grace Hopper
> Dame Mary Lucy Cartwright   Cora Ratto de Sadosky
> Stephen C. Kleene   Nicolas Bourbaki

Remember to use all of the resources at your disposal. Begin with *Yahoo*! and then move to the other search engines. Check the obscure, as well as, the popular resources. You are also welcome to read ahead to Chapter 3, which explains how to use e-mail and newsgroup resources. You never know what you'll find.

Chapter 3
Doing Your Own Tune-up

By now you have probably wandered and drifted through some of the many resources offered by the Web. In this chapter we will cover additional techniques to help you make your Web experiences more enjoyable and beneficial.

Section 3.1
Fuzzy dice and seat covers

On the Internet, as in the rest of life, time is money. Saving time is critical if you have to use a modem and an Internet service provider (ISP) to cruise the Net. If you know what your browser can and cannot do, and if you configure the options for your browser so that it is most efficient for you, you will save oodles of time and money. Try these modifications described below as you read; also, feel free to jump around to the information that is most important to you.

With *Netscape Navigator*, you have several ways to customize the appearance and behavior of your browser. Some of you may have experimented with these options already, but here is a description for those who haven't. These features are known as *Options*. Descriptions for some of the *Options* features follow, as well as reasons you might want to change them. After you get the hang of *Options*, you'll be able to explore the other custom features on your own.

> Some of you may want a more detailed description of your *Netscape Navigator* browser. *Netscape Navigator* has included access to a detailed set of descriptions and instructions on its homepage. To access the *Navigator Handbook*, select *Help* from the menu bar and choose *Handbook* from the menu list. This will connect you to Netscape's online handbook. Because this information is online and not held within your browser, you need an active Internet connection to use it. You can access this same information from the following URL:
>
> http://home.netscape.com/eng/mozilla/3.0/handbook/

Getting Options
Netscape Navigator lists common functions under the *Options* menu on the menu bar at the top of the window. Select *Options* and a pull-down window will appear (Figure 3.1). Notice that there are check marks by some of the options. Look at *Auto Load Images* for example. When the check is present, images will be automatically loaded. If you select

this option again, the check mark disappears and subsequently, pages will load the text without the time-consuming images. Give it a try. *Netscape Navigator* lists only the basic options here. Experiment with the other basic options (each of these begin with *Show*) to see what they do.

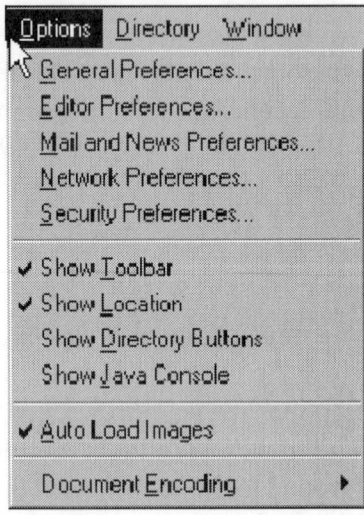

Figure 3.1. For *Netscape Navigator*, many of the general options are listed below the *Options* menu. If you wish to do a more detailed configuration, you can select the different *Preferences* available at the top of the box.

More detailed options can be accessed from one of the *Preference...* sub-menu options. (You can see these at the top of the *Options* pull-down menu.) Once you've selected one of *Navigator's Preferences...* windows, you'll notice that the preferences are broken down into categories according to function (Figure 3.2). The different categories are listed at the top of the window; they look like tabs in a file folder. As I describe different customizing options, I will refer to both the specific preference window and the specific tab within that window (i.e. *General Preferences...* window and *Appearance* tab).

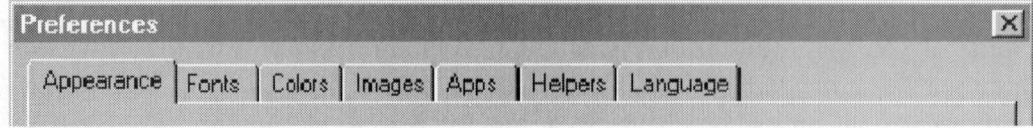

Figure 3.2. These are the *General Preference* categories in *Netscape Navigator*. This illustration includes only the categories and not the contents of the window. The contents will vary depending on the tab selection. A different set of tabs and categories exist for each of the different *Preference* windows.

Now that you know where to go to modify the setup for your browsers, I'll describe some of the features you may want to consider modifying. (Remember, you don't have to make all of the changes described. Change your options if the modifications make sense for you.) You will probably want to look at each preference even if you choose not to modify it. Also, keep in mind that some of these changes are not relevant to your setup; others require specific information so that your browser will work with your ISP. You can always check the online *Navigator Handbook* for help.

28

Homepage

You might want to modify the location of your homepage. A homepage is sort of like home plate in baseball, or the garage where you always take your car for a tune-up. If you're working with a browser that hasn't been customized before, the company that made the browser probably chose the homepage. If you're working on one of your school's computers, then the homepage may already be set to the school's page. Find out if you are permitted to change the homepage in your computer lab. If it's okay to do so, then decide what you want to set as your homepage. By now you've probably found something on the Web that you are willing to call home. When your homepage is properly set, every time you start your browser or select home from the toolbar, you'll end up at this place, so choose wisely. (However, you can always change your homepage again if you want to.) Later in this chapter, we'll discuss how you can make and display your own homepage. Stay tuned.

In *Netscape Navigator*, open the *Options* menu, choose the *General Preferences* window and then select the *Appearance* tab. At the center of the window you will see a dialog box within the *Startup* control set (Figure 3.3) that allows you to set your homepage. Type in the URL of your new homepage.

Figure 3.3. On *Netscape Navigator*, the *Startup* control set within the *Appearance* tab of the *General Preferences* includes a dialog box that will allow you to enter the URL for your desired homepage.

Link Age

You have probably noticed that hyperlinks are normally blue (unless someone has changed the default color). Once you've accessed the page that the hyperlink represents, it changes to purple. Think of this as the bread crumb principle. The purple links remind you that you've already been down a particular path (link) and that you might want to select a different one. Eventually, however, the hyperlink will change back to its default color. But you can determine how much time will elapse before a hyperlink changes back to the default color. You can also choose the default color. Generally, the more time you spend on the Internet, the more quickly your links should revert to default. I have my link expiration set to one day because I do so much traveling. If you travel frequently and find that your hyperlinks are always purple, then your expiration limit may be set too long.

From the *Options* menu, open the *General Preferences* window and select the *Appearance* tab. At the bottom of the window, you will see a dialog box that allows you to manipulate the length of time in days for hyperlinks expiration (Figure 3.4).

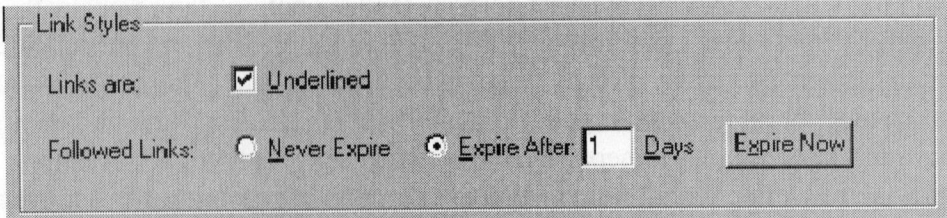

Figure 3.4. On *Netscape Navigator*, the *Link Styles* control set within the *General Preferences* tab group allows you to control the appearance and behavior of hyperlinks.

Font Style, Size, and Color

If you spend a great deal of time on the computer, then you realize the importance of adjusting your monitor, desk, and keyboard to minimize potential stress. Eye fatigue and/or muscle tension are possible symptoms of a poor setup. Selecting a font of appropriate color, style, and size will also help you prevent both eye strain and muscle fatigue.

To change the default style and size of the font, select the *Options* menu and open the *General Preferences* window. Select the *Font* tab from this window (Figure 3.5). You may modify the default settings for both the font size and style used by your browser.

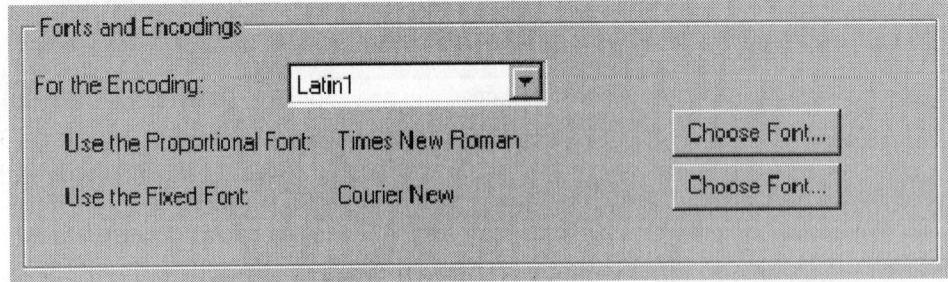

Figure 3.5. On *Netscape Navigator*, the *Fonts* tab group within the *General Preferences* option window allows you to control the default font size and style of text elements.

To view other possible font settings, you need to select the appropriate *Choose Font...* button. You should work primarily with the proportional font setting because it is used for the normal text on your browser. (Fixed fonts are generally used when a page designer is illustrating computer code or text requiring even spacing, which does not occur very often.) A small window will appear; in it you can select from numerous font styles and sizes. When you select the *OK* button, those font settings will be selected as the default for your browser. If you like your selection, then you need to again select the *OK* button at the bottom of the Fonts tab window.

To change the default color of the font, select the *Options* menu and open the *General Preferences* window. Select the *Colors* tab from this window (Figure 3.6). You may modify the default settings for font color used by your browser for the hyperlinks (fresh and followed) and regular text.

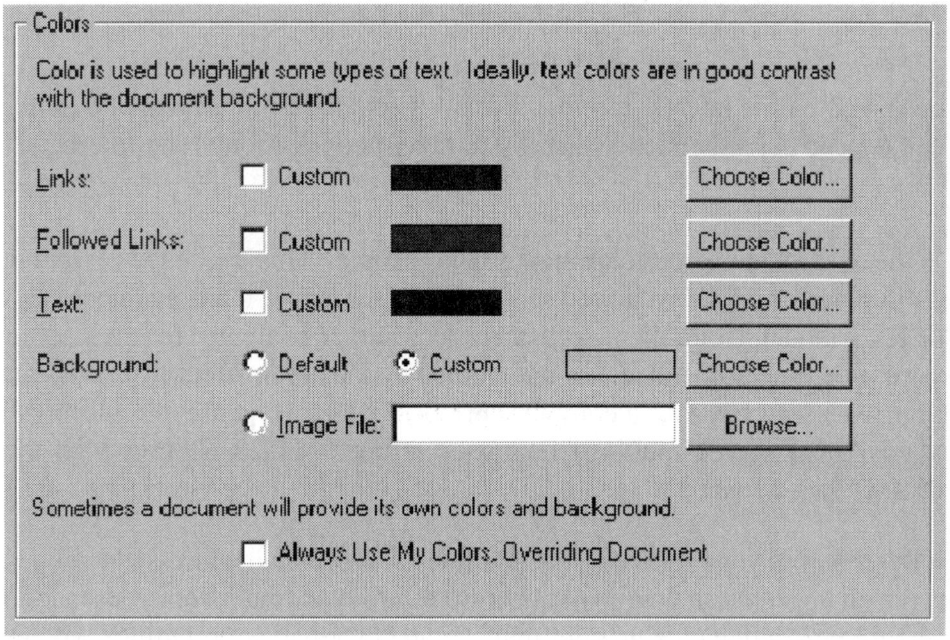

Figure 3.6. On *Netscape Navigator*, the *Colors* control set within the *General Preferences* option group allows you to control the default color of the different text elements.

To change the color setting you need to select the appropriate *Choose Color...* button. A small window will appear displaying numerous color choices. When you select the *OK* button, that color will be selected in the *Colors* tab window. If you like your selection, then you need to select the *OK* button from the *Colors* tab window.

Experiment with the *Colors* tab window a while. Try to manipulate both the text and background color. Also, try to discover how you might use an image in the background instead of a solid color. Remember not to go overboard with your power; you still need to be able to read the words on the page. Some font colors or sizes may be difficult to read; these will cause more eye strain.

Although we have not discussed all of the options, you probably have enough experience to understand how to modify them. Experiment with the settings on your own. However, you do need to be careful. Some of these options won't apply to your setup; with others, you will need specific information to ensure that they will work properly with your ISP. You can always check the online *Navigator Handbook* for help. If you are using *Microsoft's Internet Explorer* or another browser, you will find that they have very similar features.

It doesn't matter how you customize your browser as long as you're productive and happy (or even just happy). All I have to say is, *To thine own self be true.*

Section 3.2
Horns, radios, and turn signals

Netscape Navigator and most other browsers allow you to communicate through e-mail, newsgroups, and online interest groups. E-mail links are built into many Web pages and enable you to send correspondence directly to other people through the Internet.

E-mail
E-mail is the electronic exchange of mail among people. This exchange is from one person to another person. Like web browsers and web servers, e-mail operates according to a Client-Server model. Your client software asks a server computer for mail addressed to you. Your mail server is operated and maintained by the administrator of your campus or ISP. If your Web server speaks a computer protocol called HyperText Transport Protocol (HTTP), your mail server speaks a computer protocol called Simple Mail Transport Protocol (SMTP).

Think of the server as your mailbox. The mailbox is where mail addressed to you is stored until you pick it up. You use your mail client to retrieve your mail from your mailbox. (Be careful about forgetting to pick up your mail. If you overload your mailbox, you may lose some pieces.)

Access to an e-mail server is easily available. If you arc at a campus that has provided you with Internet access, you should be able to apply for an e-mail account through your campus computer administration. Otherwise, you can apply for an e-mail account through an ISP. If you use an ISP to access to the Internet, you probably have e-mail capabilities. In order to properly configure your browser for sending e-mail, you need to know (1) your e-mail address and (2) the name of your mail server.

Here's a little bit on why e-mail addresses look the way they do. The format of a typical e-mail address is as follows: NAME@HOST.DOMAIN. It is not necessary to have a full name for the NAME part of the address; in fact, some addresses use only numbers to represent an individual.

> Here is an example of a typical e-mail address.
>
> **andrew_stull@prenhall.com**
>
> A typical e-mail address has three basic components:

user name	andrew stull
host server	prenhall
domain	com

This is my e-mail address. Please drop me a brief note if you have some suggestions for making this manual better. Your comments and suggestions are important.

Your *User Name* is also the NAME associated with your e-mail address. The @ symbol always follows the NAME and then the name of the server computer (HOST). The domain in the e-mail format, just like the domain of the URL format, is used to indicate the affiliation of the user. Notice that there are no spaces anywhere in an e-mail address.

Your campus computer administrator or your ISP will provide you with an e-mail address. This is the address you'll give to your friends. It will also be posted with any correspondence you send on the Internet.

Configuring your browser to send and receive e-mail only takes two steps.

<u>First</u>: To set your personal information, select the *Options* menu and open the *Mail and News Preferences* window. Select the *Identity* tab and you will see the text entry boxes where you can enter your information (Figure 3.7).

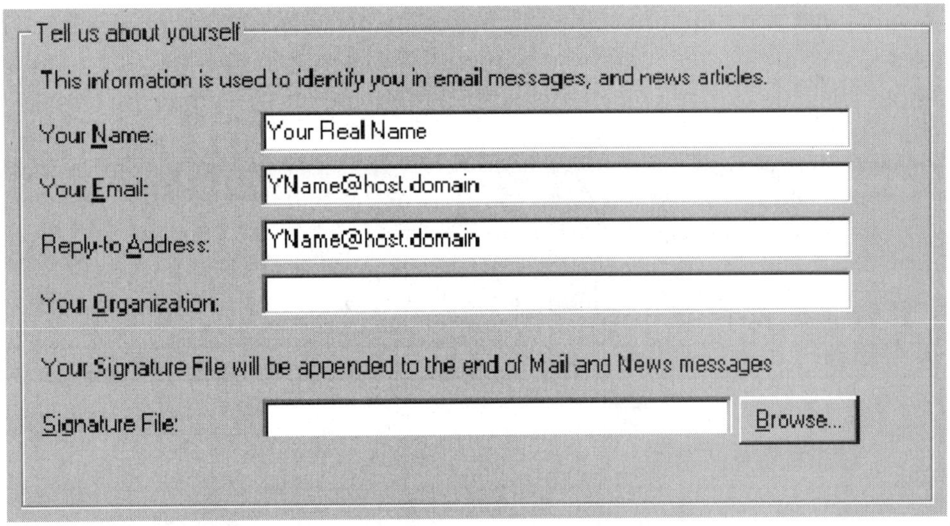

Figure 3.7. This information helps to identify you when you send e-mail. Use the *Signature* feature to place standard information (such as your address and phone number) at the bottom of your messages.

Second: To set your e-mail information, select the *Options* menu and open the *Mail* and *News Preferences* window. Select the *Servers* tab and you will see the text entry boxes (Figure 3.8). After you've entered the correct information, restart the browser program. Now you should have the ability to send and receive e-mail. Think about it! You'll save on postage and long-distance phone bills.

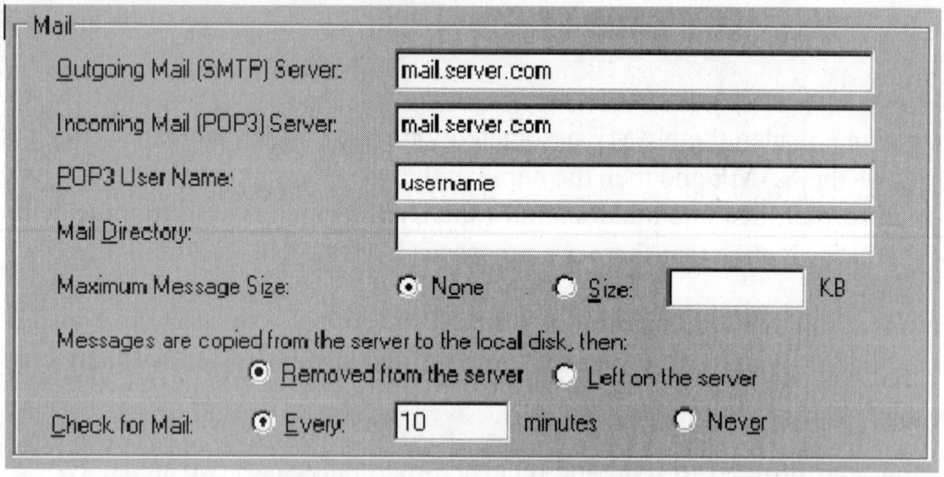

Figure 3.8. The information about your particular mail server and e-mail address must be entered before you have the ability to receive e-mail from others through *Netscape Navigator*.

Most companies use a single machine to handle both incoming and outgoing mail. However, some e-mail systems use different machines for incoming and outgoing mail. In Figure 3.8, note that two different text entry boxes are provided for the mail server. If your system uses just one machine, enter the mail server twice. The person who sets up your account should be able to give you all the information you need. If they are less than help-

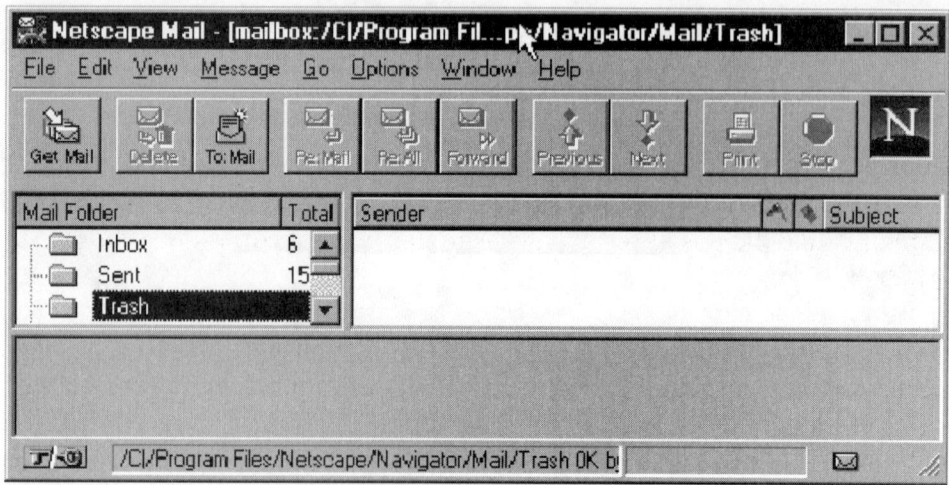

Figure 3.9. *Netscape Navigator* includes a built-in mail client. Its features are similar to those of a Web browser.

34

ful, then it is generally safe to assume that the address for both your incoming and outgoing mail server is MAIL.HOST.DOMAIN.

Now, let's send some e-mail. Select the Windows option from the menu bar and choose *Netscape Mail*. A new window will open (Figure 3.9).

Notice that this window is similar to but slightly different from a Web browser. (It is also similar to the news reader that we will discuss later.) To view your mail, select the *Get Mail* button. If you have mail, then it will appear on the left side of the window. All incoming mail arrives in the *Inbox*. When you select the *Inbox*, its contents will appear on the right side of the window. Finally, when you select a specific mail item you will see the contents displayed at the bottom of the window. Take time and explore, but remember you've got to send mail to get mail. Maybe you know someone else with an e-mail address. Send them a note.

> Now that you've got the basics down, here is a site that provides more information on e-mail. It was written by Kaitlin Duck Sherwood.
>
> **A Beginner's Guide to Effective Email**
>   http://www.webfoot.com/advice/email.top.html

Newsgroups

Many browsers enable users to exchange ideas by using *newsgroups*. A newsgroup is a group of people that connect to and participate in a specialized discussion. These groups are open forums in which all are welcomed to contribute. Some are moderated by an individual or individuals who post items for discussion and moderate any electronic brawls that may ensue. Others are free-form and function much more like a street fight. All newsgroups provide a place where people can bring new ideas and perspectives to the table. However, not all newsgroups are rife with disagreement. Many offer excellent opportunities for polite conversation with courteous people.

Newsgroups are nearly the opposite of e-mail. Where e-mail is sent and posted for a single person to read, news postings are sent to a common place for all to read and comment on. If e-mail is like the postal service, newsgroups are like coffee houses on open microphone night.

What is the purpose of the newsgroups if people just contribute ideas and comments to a mass discussion on a particular topic? Newsgroups are a place where you can go and ask questions, get ideas, and learn about specific topics. Select a group that has an interest common to yours or a topic that you are trying to learn something about. You'll find that there are many newsgroups that focus on issues of science, as well as anything else you could think of.

The names of newsgroups usually describe their discussion topic. There are several major newsgroup categories. The one you'll use most is the SCI group. This group contains most of the general science discussions. Other groups include COM for computer topics, and ALT for alternative topics (this is where you find most of the really interesting newsgroups - ALT.SCI.TIME-TRAVEL, for example). A list of science- and mathematics-related newsgroups is included in the appendix. You'll find some of them interesting and helpful as a supplement to your studies in mathematics.

> Here is a typical newsgroup URL. Newsgroup names are divided into several descriptive words separated by a period and organized according to a hierarchy. Note that newsgroup URLs don't use the double slash (//).
>
> **news:sci.math.symbolic**
>
> Here are the three components:
>
> **protocol**   news:
>   **top category**   sci
>   **sub category**   math
>   **sub category**   symbolic

To configure your browser to connect to a newsgroup, you will need the name of the server computer that handles newsgroups. (Get this from the same people who gave you your mail server address.) This is almost the same procedure you performed to set up your e-mail service. To set your newsgroup information, select the *Options* menu and open the *Mail* and *News Preferences* window. Select the *Servers* tab and you will see the text entry boxes.(Figure 3.10). Once you enter the correct information and restart the browser program, you should be able to view newsgroups. You may find that you can't access some newsgroups. Check to make sure your ISP subscribes to these. If not, they will usually do so if you make a request.

News servers use a protocol called *NetNews Transfer Protocol*, or *NNTP*. You might need to ask around to find out the address for your news server. If you can't find someone to help, you can probably assume that the address for your news server is NEWS.HOST.DOMAIN.

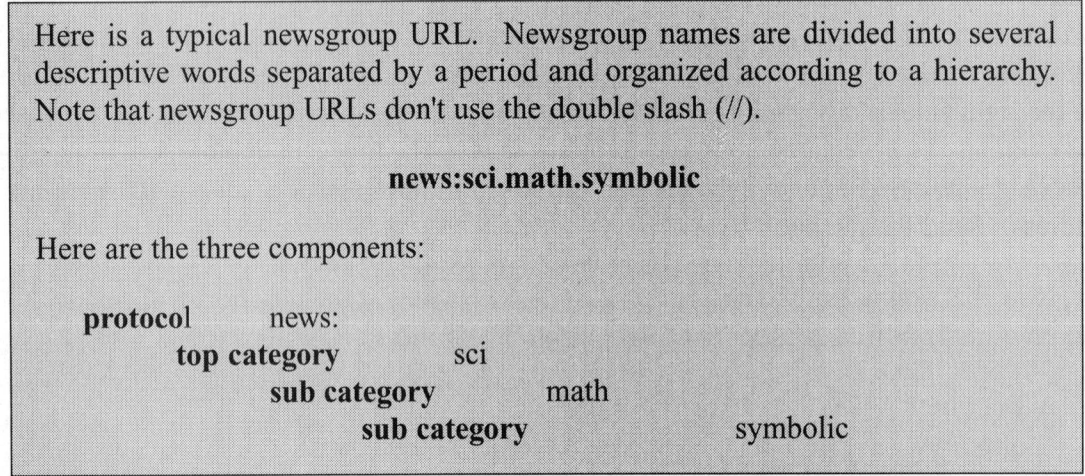

Figure 3.10. With *Netscape Navigator*, the dialog window where you will enter the name of your news server can be found in *Mail* and *News Preferences…* option under the *Options* menu.

You should now be ready for an adventure. When you use *Netscape Navigator* to read or post articles to newsgroups, you will be using a different window than what you use to browse the Web. This window is typically called a news reader. To start your news reader, select the Windows option from the menu bar and choose *Netscape News*. A new window will open (Figure 3.11).

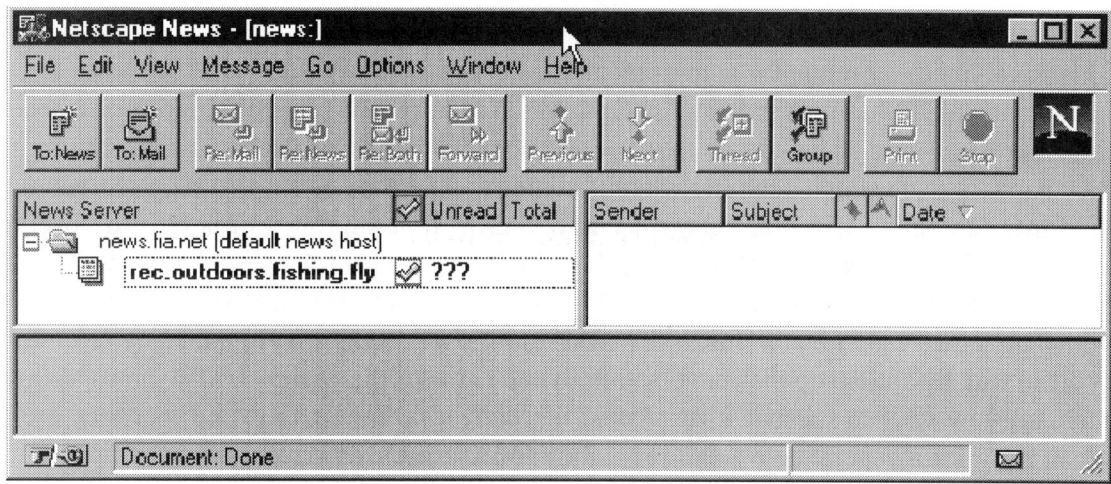

Figure 3.11. *Netscape Navigator* includes a built-in news reader. Its features are similar to both the mail client and the Web browser.

Note how this window resembles the Web browser, as well as how it differs. To view a larger list of newsgroups you can select *Options* from the menu bar and choose *Show All Newsgroups*. A larger list of newsgroups will appear on the left. You can subscribe to a specific one by placing a check mark by the desired newsgroup. When you select a newsgroup, the individual articles will appear on the right side of the window. Finally, when you select a specific article, you will see the contents displayed at the bottom of the news reader window. Now, take some time and explore.

> The only way to really understand what newsgroups are like is to try them. Here are a couple of newsgroups that will get you started. Connect to one of interest and read some of the postings. You can find a larger list in the appendix. At first, you should just read the postings and follow some of the conversations. (This is called lurking.) Connections or threads develop between postings as people add comments to earlier postings. As you become familiar with the topic of discussion, you might post a comment yourself.
>
> **news:geometry.puzzles**
> **news:sci.math**
> **news:alt.math.undergrad**

> To get the real flavor of a newsgroup you need to spend more than just a couple of minutes viewing one. Choose a newsgroup of interest. Within that newsgroup, select a topic that is being discussed and follow it for a week. As a discussion grows, you'll notice that many people will jump in to participate. Before you jump in, you may want to read the section titled "Language for the Road."

Chat

This is yet another avenue for communication. It is probably the one you'll choose if you like to gab. Unlike e-mail and newsgroups, which require you to wait for a response, chat rooms are real-time. This means that you are conversing and observing conversations as they happen. Now, I'm a beginner when it comes to chat rooms, but I've received enough help from an online friend to get you started. As with newsgroups, it is best to observe the interactions of several chat rooms and read the new user information before you jump in. Some rooms can be excessive and vulgar, but many are frequented by polite people with a genuine desire to gab. Just like newsgroups, chat rooms are organized by topic and you can usually anticipate the conversation by the name of the group. Unlike newsgroups, there are few science-specific chat rooms; therefore, you'll probably find that they are a great place for a general conversation but not necessarily for a serious conversation.

> Just to get you started, here are a few Web-based chat groups. Some of these come to me by way of a very kind friend currently at Temple University.
>
> **WebChat Broadcasting System**
>   http://webchat5.wbs.net/
> **Jammin's ChatPlanet Chat Room!**
>   http://www.jammin.com/livechat.html
> **Matt's PaleoChat Room**
>   http://www.pitt.edu/~mattf/PaleoChat.html
>
> Read the instructions and ask for help if you get confused.

Unlike e-mail and newsgroups, you will be frequently asked to log-in or register before you can chat. Often you'll even be given a password. Always, read the rules for participating and ask for help if you don't understand something. The participants are generally very helpful. The other character of chat rooms is that typically you will use a handle or pseudonym when you post a message. Just like anything else, chat rooms are not for everyone. Some find them to be more entertaining than educational, but decide for yourself.

Language for the Road

A few words of caution. Unlike normal conversations between people, electronic exchanges restrict your ability to use vocal inflections, facial expression, and body language while speaking. (These things aren't easily digitized and transmitted across a copper wire unless you have an Internet telephone or video conference. These things are available and worthwhile if you have a powerful system, but for most of us they are not practical.) You will need to practice the way you communicate electronically and be patient when someone misinterprets something you've "said." For example, suppose you tell a joke. Because the reader only gets the text, she may think that you are serious and take offense. Computer hacks learned about this problem a long time ago and several innovative solutions have been developed. They are called smileys.

Here are a few—you'll have to turn your head side-ways to view them:

:-)	This is a happy face
;-)	This is a wink.
:-(	This is a sad face.
:-[	This is a really sad face.

Because you eventually will get tired of typing all of the things you have to say, several acronyms are in common use. Here are but a few:

BTW	By The Way
IMHO	In My Humble Opinion
FYI	For Your Information
FAQ	Frequently Asked Question

Now for a couple of terms. Like any other culture, the Internet has borrowed some common words but use them in a context you might not normally expect. Here are two that you'll commonly see:

Flame This refers to the act of yelling, insulting, or degrading a person or his or her character. You can expect to get flamed if you don't follow certain basic rules of *netiquette* (etiquette of the Net).

Spam This refers to the act of posting a comment, message, or advertisement to multiple newsgroups when the note doesn't really pertain to the newsgroups. It is the Internet equivalent of sending junk mail.

These are a few Web resources that you will definitely want to read when you begin communicating online. They are very helpful and at times even pretty funny. Enjoy!

E-Mail Etiquette
   http://www.iwillfollow.com/email.htm

> **The Net: User Guidelines and Netiquette By Arlene H. Rinaldi**
>   http://rs6000.adm.fau.edu/rinaldi/net/index.htm
> **Electronic Frontier Foundation's Unofficial Smiley Dictionary**
>   http://www.eff.org/papers/eegtti/eeg_286.html

So, you displayed good netiquette, you use the smileys, you're aware of the way you tell a joke, and you still get in trouble. Well, *C'est La Cyber Vie.* There are a great many people on the Web and some think that because they have anonymity they can misbehave. Others simply react before they think about what they are going to say. Again, without the immediacy afforded by a face-to-face conversation, people may misinterpret what you say. Be patient, give them the benefit of the doubt, always offer a kind word, and if you realize you were the one with the quick word, apologize. It will be well received. Eventually, you'll meet some really nice people online.

Section 3.3
Turbo chargers

In early Web history, travel was pretty basic. The browser was a piece of software that acted as a totally self-contained vehicle. When you hooked it up, it contained everything you needed to cruise the Net. This has changed. As the Web grows larger, people develop new formats of information they wish to include on the Web. These new formats allow for a richer experience. Things like digital video, audio, animations, interactive games, and 3D worlds are currently very common on the Internet. Eventually, as people demand more functionality from the Internet, new and richer formats will be developed and available. To cope with this continuing change, *Netscape Navigator* developed a way to add functionality without overloading the basic browser design.

Helpers and Plug-ins
Early Web pages included only text and images. They were much like brochures with the ability to hyperlink to other pages. The images that were included in Web pages were in a GIF format. Although other image formats were available, the standard was set to GIF. But, as you may have guessed, some people wanted another image format, JPEG. (The difference between these two formats is not important to our discussion.) To accommodate this new format, browsers turned to helper applications. If a page designer included a JPEG image, then an external program, a helper application, was used to display the image outside of the browser (so you could view it). The next step was obvious. Instead of using an external program, why not allow computer people to develop programs that could (if the user plugged them in) add new features to the basic browser? This is the difference between a helper and a plug-in. A plug-in is a program that is added to the internal workings of a browser. For example, if you wanted to listen to music across the Internet from your browser, then you could install a plug-in that would allow the browser

to understand the audio format. So, a plug-in is not an external program but rather an internal extension of, or addition to, the basic browser.

Netscape Navigator was critical to the development of many plug-ins, and they are still the place to turn when you want one. One important thing to remember—adding plug-ins to your browser will increase its requirement for memory (RAM). If you go overboard with the plug-ins, your browser might stop working. Only add the plug-ins that you're going to use regularly.

> These addresses will take you to the place where you can download the plug-ins to your computer. These developers also provide a very good description of how to install them and make them work for you.
>
> **RealAudio by Progressive Networks**
>   http://get.real.com/products/player/download.html
> **Shockwave by MacroMedia**
>   http://www.macromedia.com/shockwave/download/
> **QuickTime by Apple**
>   http://www.quickTime.apple.com/sw/
> **Live3D by Netscape**
>   http://home.netscape.com/comprod/products/navigator/live3d/

Once you've lived a little and added a plug-in or two (or too many), you can get a list of the ones you've installed right from your browser. Select the *Help* menu and then choose *About Plug-ins*. The browser window will then present a list of all of your installed plug-ins. If you click on the link at the top of the page, you will also be able to connect directly to Netscape's master plug-in page.

JavaScript and Java

Grab a cup of coffee for this one. I'm going to lead you down a twisty path, but I promise the view is worth the potholes you'll have to dodge. As you'll learn shortly, Helpers and Plug-ins weren't enough. Although Helpers helped and Plug-ins plugged, Web developers wanted even more—*more* flexibility, *more* interactivity, and *more* control. *JavaScript* and *Java* were developed to provide the *more* that developers wanted. But to understand what they are, you need to understand what Web pages *are not*. (Here is where you have to watch the road, it gets bumpy.) Web page designers don't claim to *program* Web pages; rather, they say that they *write, design*, or *create* Web pages. So, you may wonder what the difference is. Programming is the act of developing a set of commands that perform a specific function. Writing, designing, and creating are acts that generate things that communicate ideas. You may still be confused but consider this. *Programs actively do things* (sometimes on their own and sometimes because you command them to), whereas *Web pages communicate ideas* (by bringing together different media). One type of media, interactive programs, did not exist until Java and JavaScript were made available. They are both programming languages that are used by people to write programs that do things on Web pages. They add functionality to Web pages just as helpers and plug-ins do, but they provide much greater flexibility. The fundamental difference between the two is that JavaScript is usually integrated with the Web page and runs on the client side, whereas Java is usually integrated on the server side. The differences are probably best understood by examining the programs for yourself. If you are using Netscape 2.0 or greater (or Microsoft Internet Explorer 3.0 or greater), then you have the ability to view and interact with these programs.

> The first three URLs will connect you to some really interesting examples of Java.
>
> **The Impressionist**
>   http://reality.sgi.com/employees/paul_asd/impression/index.html
> **Crossword Puzzle**
>   http://home.netscape.com/comprod/products/navigator/version_2.0
>   /java_applets/Crossword
> **TicTacToe**
>   http://www.javasoft.com:80/applets/TicTacToe/example1.html
>
> This last URL will connect you to a rather large collection of both Java and JavaScript examples.
>
> **Gamelan**
>   http://www-b.gamelan.com/index.shtml
> Depending on your system, these examples may take a while to load so be patient.

Activity: A Talkabout Tour

I'm sure you understand that learning doesn't stop when you leave the classroom or laboratory. Important resources to help you with your education also reside in textbooks, libraries, and on the Internet. Let's extend this line of thought. Have you yet noticed that your instructors are finely-tuned, cross-referenced, interactive information storage and retrieval tools? (If they talk with their hands, we could even say they were multimedia.) Everyone brings a unique perspective to the table in a discussion. That's why newsgroups and chat rooms are so popular. In the same way that your instructor is a resource for information, you and the people around you are important information resources. Take the time to look away from your book occasionally and you'll discover some wonderful things.

E-mail
Because you have a wealth of information at your fingertips, take the time to get some e-mail addresses from the members of your class. Most good e-mail software will provide you with a mechanism to store these addresses for easy retrieval. Begin exchanging e-mail with some of these people and you'll find that there is plenty of help available to you outside of class.

Newsgroups
Make it a practice to browse newsgroups. Plenty are of general interest; you should also make a habit of checking into a newsgroup that focuses on important topics from your class. By visiting many different newsgroups, you'll always have the latest scoop. You also have a resource that you can tap for help.

Chat
Chat rooms may or may not enhance your education. However, you might be able to find a chat room that is relatively empty. If so, arrange a fixed weekly time when you and your classmates can get together to work out problems or discuss material from that week. A better option is to meet in person, but if this is impossible, chat rooms can provide an alternative.

Web Page
If you are using an ISP for Internet access, then you probably have access to posting information on your homepage. Consider using it as a way to bring other students together for discussion. Check with your instructor and you may find that he or she is willing to help you with this activity.

If you've been able to do even a little of what we've discussed here, you'll realize how connected you are with the world.

Chapter 4
Hot Rodding

Web sites are like cars in that some people simply enjoy using them, whereas others derive pleasure from building and maintaining them. You don't need to know what's "under the hood" in order to appreciate the Internet. However, if you're curious, confident and want a little bit of extra knowledge, then this chapter will help you understand how to design a Web page.

To gain the most from this chapter, you will need a few basic skills. You should be very familiar with the operating system on your computer, capable of navigating its directory structure and competently creating, moving, and renaming both its directories and files; and able to install applications onto your computer and set them up to function properly. If you are unfamiliar with any of these tasks, you may want to consult with other students, computer technicians on campus, and your user's manual for an overview. With these skills, you should be able to follow the concepts outlined in the rest of this chapter. Keep in mind that what follows is not a step-by-step description of Web design, but an introduction to some basic concepts. You may want to supplement this discussion by consulting additional resources.

Section 4.1
There's no place like home

There *is* no place like home, or your very own homepage. The following description will help you create one for yourself, if you'd like. It is not as simple as cruising the *Web*, but it's not very hard either. Web pages are stored on a Web server. However, you can create a file on your home computer, or even on a floppy disk, that will act as a homepage. Ultimately, you will need access to a Web server to make your homepage part of the Internet. If your campus has a Web server, then they may have a place for student homepages. If you are using an ISP, you will probably be able to post a homepage with them at no additional cost. For the moment, though, let's focus on building a homepage—you don't need to be able to post a homepage in order to understand how to put one together.

Homepages

What good is a homepage? You will find many uses; just a few are mentioned here. A homepage can be used to organize important and frequently accessed links and bookmarks to the Internet, keeping you connected to science and biology resources all over the world. It's better than a library card.

A homepage can also help you follow personal interests. For example, I keep links to online magazines, the weather at my favorite camping spots, movie, music, and book reviews, newsgroups about fly fishing, the current news reports, pages with online games, my school's homepage, my friends' pages (only the cool ones), the manuals for my e-mail and Web browsers, and my favorite search engines and directories. Some homepages reflect the personality of the owner, who may wish to post information about herself or himself for others to view.

Or imagine designing a homepage and turning it in as homework. You could turn in an assignment with online data, animations, video clips, and sound. You could have graphs that link directly to the data, pictures in color with an audio caption, interactive diagrams to illustrate your point, and an e-mail link if the instructor wanted to ask you a question. Move over typewriter!

Now, where is your homepage? Actually, I've already made one for you, but I'll give you a chance to make your own. If you look at Appendix II, you'll see my version. It is written in HTML, a simple language that Web browsers use to read hypertext documents. The initials stand for HyperText Markup Language. Right now, you might want to glance at it just to see what HTML looks like. Later on, you will be able to make use of this template and even liven it up with customized features.

> To get an idea of what homepages are like, use the following URLs to see what others have done. Some of these may be appealing to you, others may not—homepage design is a personal choice.
>
> **Make A Friend On The Web**
>   http://albrecht.ecn.purdue.edu/~jbradfor/homepage/

Just like with cars, you can go manual or automatic when you design a web page. Both methods have advantages and disadvantages. When you design a page manually, you actually write with HTML (like shifting gears for the car). You'll have to learn some rather obscure notations, but you will be able to fully control the look and feel of the page you create. Because most people want an easy way of doing page design, tools have been developed to automate the process. These tools are called WYSIWYG (pronounced wizzy-wig) editors. (WYSIWYG stands for What You See Is What You Get.) The disadvantage to WYSIWYG tools is that you have less control over page design, but you also save a considerable amount of time. Many designers use WYSIWYG tools for their initial layout work and then tweak the HTML code to get what they want. Here is a little bit about each method. It should be enough to get you started.

Automatic
You can use several tools for WYSIWYG design. Some of them offer you the opportunity to view the HTML as you design your page; in others, the code is hidden. We will focus

on Netscape's WYSIWYG tool, which is called *Navigator Gold*. Basically, it consists of Navigator with a built-in editor. (I'll refer to these two parts as the Browser and the Editor.) This software is free to both students and educators and you can get it directly from *Netscape Navigator* across the Internet. Before you read on, download *Navigator Gold 3.0* from Netscape's download site and install it on your computer (http://home.netscape.com/comprod/products/navigator/gold/index.html).

> Choice is a great thing—exercise yours! Instead of relying solely on the information people give you, check for yourself, too. You may turn up something that meets your needs better. In this light, here are a few more WYSIWYG editors.
>
> **Microsoft FrontPage**
> http://www.microsoft.com/frontpage/
> **HoTMetaL Pro**
> http://www.sq.com/products/hotmetal/hmp-org.htm
> **HotDog Pro**
> http://www.sausage.com/
>
> Investigate the numerous WYSIWYG editors on the market carefully to make sure you will be able to create the homepage you want. Henry Ford said that when you bought one of his Model A's, you could have any color you wanted as long as it was black. Some WYSIWYG editors are the same way.

Now, we're ready to get started. You should have *Netscape Navigator Gold 3.0* installed and an active Internet connection. If you select *Options* from the menu bar, you will see an additional preference command titled *Editor Preferences…*. Select this and a window will appear (Figure 4.1). Three tabs are found at the top of this preference window. Select the *General tab* and enter your name (as you will soon be an author). You should also enter information for your preferred HTML text editor and image editor.

Are you wondering why you need another HTML editor? Keep in mind that WYSIWYG editors are not the end-all and be-all of document creation. Occasionally you'll want some manual control over both the pages and the images you embed in them. But you don't need to buy any more software. All you need is a simple text editor, which is probably included with your computer. If you are using a Macintosh operating system, then you can use either TeachText or SimpleText. If you are using a Window-based operating system, then you can use either Notepad or Wordpad. Use the *Browse…* button to locate the text editor you'd like to use.

The image editor may not be included with your operating system. However, you don't need one to follow our discussion. When you need one later on, check on the Internet, where numerous image editors are available for free. Use your search engine to locate one.

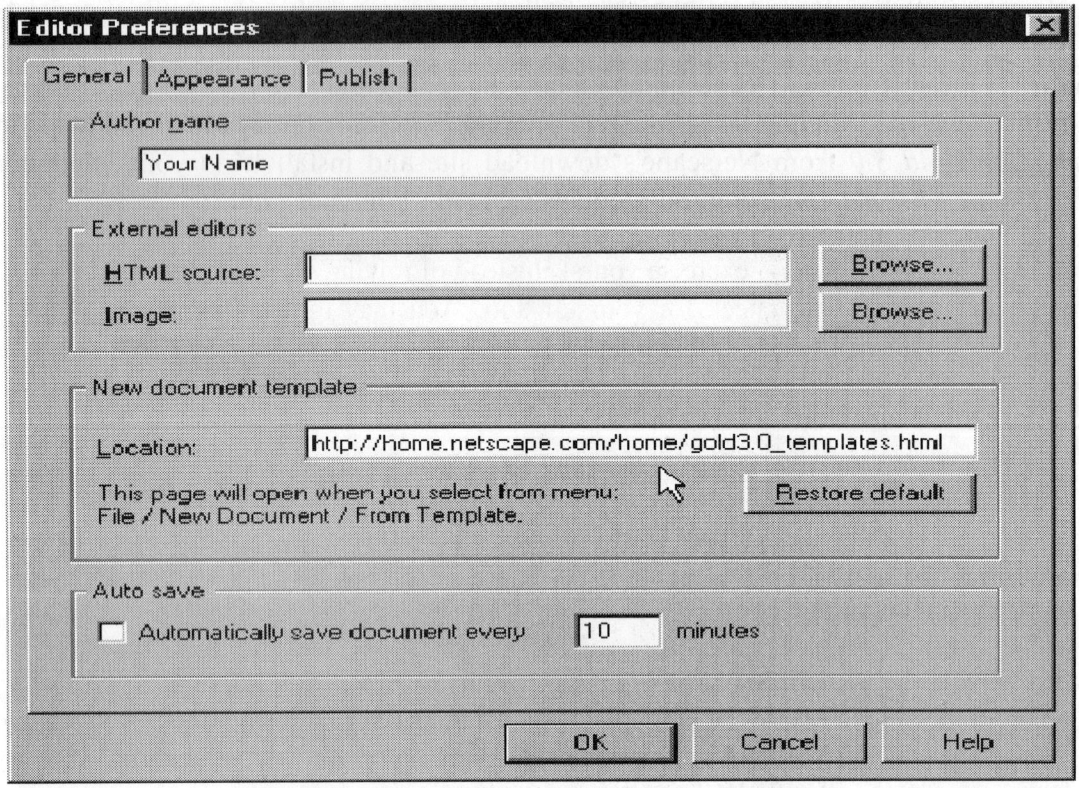

Figure 4.1. Opening the *Editor Preference* selection in *Navigator Gold* will activate this window.

The next section in the *General* tab is where you enter the URL for a page template source. Don't worry about it now. Select the *OK* button and your settings will be saved.

Templates

Netscape has organized a wonderful little resource to help first time Web authors such as yourself. With this, you will quickly and easily write your first page. This is done online and all you need to do is follow along. In no time, you'll have a simple homepage. As your skills develop further, you might even turn it into something spectacular.

To get started, select *File* from the menu bar and choose *New Document* and then choose From *Wizard*...(Figure 4.2). You may also use the following URL (http://home.netscape.com/home/gold3.0_wizard.html).

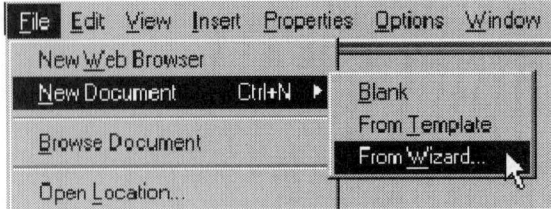

Figure 4.2. *Netscape's Page Wizard* is accessible from the menu bar on *Navigator Gold*.

48

The *Page Wizard* (Figure 4.3) is divided into three unique windows (or frames). Read the basic instructions from the upper right window and select START when you are ready.

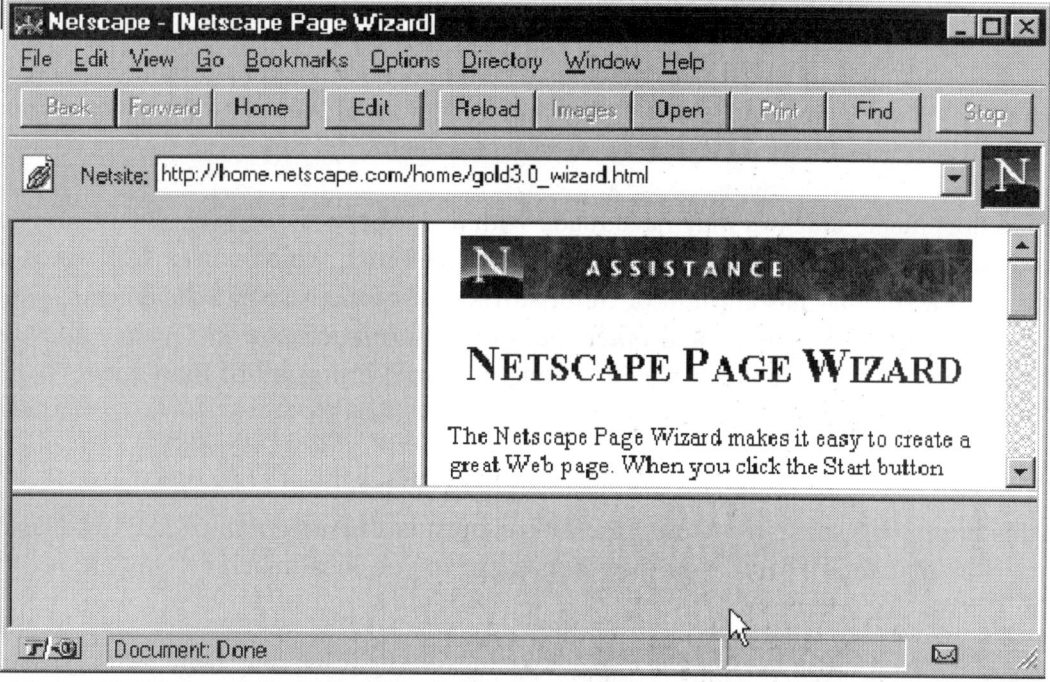

Figure 4.3. *Netscape's Page Wizard* is available only online and you will need to have an active Internet connection to use it.

Once you've started, you'll see a set of instructions in the upper left window. Although the instructions are fairly self-explanatory, a brief overview of what you need to do is provided here. You'll be working in a roughly counter-clockwise direction.

1. Read the instructions in the left window.
2. Click each hyperlink as you come to it in the instructions.
3. Enter the requested information in the lower window.
4. View the information as it appears in the right window.
5. Modify the information (lower window) until you like what you see.
6. Continue executing the instructions until you get to the end.

When you've executed all of the instructions, select the *Build* button and your first homepage will be displayed in the *Browser* window.

So—what have you accomplished so far? Essentially, you've used a remote tool (*Netscape's Page Wizard*) to create a homepage from a preset template. Although it may not seem this way, everything that you are now looking at is still on Netscape's server and not your computer. You can't do anything more to your homepage until you move it over to your machine.

Select the *Edit* button from the tool bar at the top of the *Browser*. (If you aren't using the Gold version, you will need to download it from Netscape—it's free.) An *Editor* window (Figure 4.4) will open along with a dialog box asking if you wish to bring over the associated images and preserve the links—and you do; so answer appropriately. You will then get a dialog window in which you can determine the location on your computer where you wish to save the file and images. If you expect to do a lot of Web page design, you should create a directory where you can store all of your pages.

Unlike documents that you may have made with a word processor, Web documents don't combine all of their elements into one package. The text, which consists of words and code, is in a file that has an .htm or .html suffix. All of the associated elements, which may include images, sounds, and other non-text information, are in separate files with suffixes such as .gif, .jpeg, and others. *Navigator Gold* brings all of these pieces together when you save a page to your computer.

WYSIWYG

At this point, you still haven't used the *Editor* portion of *Navigator Gold*. Let's change that. Now that you've saved your new homepage to you local machine, you are ready to give it some character.

Figure 4.4. The *Editor* window is significantly different from that of the *Browser*.

Let's review the work you've completed so far. With *Netscape Navigator Gold* installed, you used the *Page Wizard* and generated a simple homepage from a template. You downloaded that page to your server and then you opened it in the *Editor* portion of *Navigator Gold*. There are also other templates available at the Netscape site. You can save them or any other page in this same way. Be aware that some pages you'll find on the Web contain copyrighted information.

Next, let's do a couple of things so that you can see how easy it is to use this editor. Then, you will need to experiment with the software to explore all your options. Later on, you might want to check out your local bookstore to learn more about designing Web pages. You'll find a great selection of books that will help you to understand all of the idiosyncrasies of *Navigator Gold* or any other WYSIWYG editor.

To familiarize yourself with the *Editor*, move your cursor over the top of each *Editor* button and pause (but don't click the mouse button) until the label appears. Reading the labels should help you discover the function of each button.

Something important to remember. Always save your original! This means that after you've done something cool to your page, save it as a different file. By doing so, you can always retrieve the original if you realize that your change wasn't really as cool as you first thought. It will save you time.

Here are a couple of things to do to your homepage that will start you on your way.

1. Modify (size, color, style, etc.) the text on your page.
2. Enter a new paragraph of text.
3. Enter a bullet list of information.
4. Place a horizontal line.
5. Insert an image.
6. Embed a link to an e-mail address.
7. Embed a link to a URL on the Web.
8. Create a new document and then embed a link from your homepage to this new document.

Here are two great sites to help you learn more about designing a Web page. You'll find images, icons, backgrounds, and a lot of help.

Netscape Gold Authoring Guide
   http://home.netscape.com/eng/mozilla/3.0/handbook/authoring/navgold.htm
Netscape's Gold Rush Tool Chest
   http://home.netscape.com/assist/net_sites/starter/samples/index.html

Hint: To capture an image you need only to be able to see it. For Macintosh users, click on the image and just hold down a second longer than you normally would. A window will appear with save options. For Windows, use the right button and you'll get the same options window.

Now you have a homepage! Right? Well, don't worry if it still isn't perfect. You have plenty of time to make it shine. Or you may be perfectly satisfied with what you created using the *Editor*. If so, you may want to read the next section at a later date. In it, we will explain a little bit about how to take a manual approach to design by using of HTML. The extra effort may be well worthwhile if you plan on creating more complex Web sites.

Take a look at the HTML code of the page you just created. While you have the *Editor* window active, select *View* from the menu bar and then select *Edit Document Source* (Figure 4.5). This will open the test editor you designated in the *Editor Preferences* and display the HTML code for your page.

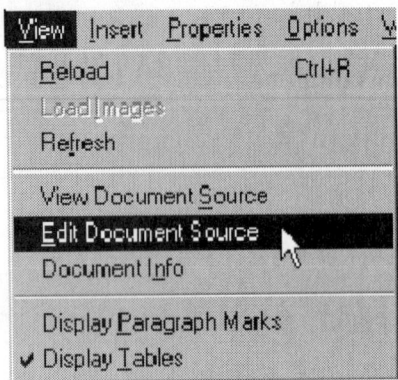

Figure 4.5. This pull-down menu is available only while the *Editor* portion of *Navigator Gold* is active. *View Document Source* will display the source but will not allow you to modify it. *Edit Document Source* will allow you to modify the HTML code.

You may want to print a copy of the HTML for your homepage. This will allow you to see it all on one sheet (unless you got carried away with the *Editor*).

Manual

The HTML code on your homepage is really pretty easy to understand. You need to remember only a couple of things in order to read HTML. For the moment, ignore the HTML and focus on the rest of the document—you'll see that it makes sense. You can do this by alternately viewing your homepage file through Navigator and then through your text editor. You don't need a fancy program to write or edit HTML, just a text editor and an understanding of the code.

Now, look at the window with the HTML code. At the very top you'll see a tag like this: <HTML>. A tag is an element of HTML code. It is added to the page to emphasize the text that will appear on the viewed Web page. Tags always have an angle bracket (<) followed by one or more letters and a reverse angle bracket (>). Here are some examples: <BODY>, <HEAD>, and <TITLE>.

Most HTML tags come in pairs. For example, the tag <HEAD>, found on the second line of the program, is accompanied by a corresponding tag </HEAD> (see the fourth line). The second member of each tag pair has a forward slash (/) preceding the word. Here are the other members of tag pairs: </BODY>, and </TITLE>. Tags, when used in sets or pairs, surround a section of text and define the beginning and end of an HTML section. It

52

is possible to use tags incorrectly just as it is possible to use English incorrectly. Here is an example of the basic HTML page structure that all properly coded pages should conform to.

    <HTML>
      <HEAD>
          This is where a description and purpose of the page is written. The material is not viewed on the browser. It is only used to document the intent of the page.
      </HEAD>
      <BODY>
          This is where the material that is to be viewed through the browser will be written.
      </BODY>
    </HTML>

All properly coded HTML documents are coded with this basic format. Notice the beginning and ending member of each tag pair. Although tags usually operate in pairs, some are used as solitary tags. When you want to separate sections of text with two spaces, you could use the paragraph tag, <P>. Two line break tags, <BR>, will also do the same thing. If you want the browser to make a horizontal line across the viewed page, then you use a horizontal line tag, <HR>. You may have noticed that the tag names are abbreviations of what they do. You don't have to learn everything about HTML now, however. You can embed some of these features with your WYSIWYG editor.

> If you really want to jump into this HTML stuff, here is a Web site that may be a helpful resource.
>
> **Netscape Page Starter Site**
>   http://home.netscape.com/home/starter.html

Other people have already written cool homepages and you might want to use theirs as a template if you don't like yours or the one in the appendix. Simply find the cool and interesting pages that other people have made. Copy the code and modify it to meet your personality.

Hyperlinks

One very important element in HTML is the code for hyperlinks (jumping points within Web pages). You know most of what you need to know to create hyperlinks. The HTML tag pair for a hyperlink is <A> and </A>. There are two main components to a link. The first element is the URL of the document to which the link leads, and the second is the word (or words) that functions as a hyperlink. You can also enter links into your pages with the *Editor*. Here is an example of how to write a link tag.

```
<A HREF="URL">
        Word
</A>
```

This may seem a little complicated, but it's not that hard. The first line of the example above begins the code for a link. The letters and signs between the angle brackets represent a URL, which tells the browser the location of the requested document. The second line is the text as it will appear on the Web page (the button, so to speak). Because it will be a hypertext link, it will appear as a colored word when viewed through a browser. The link tag pair is completed with the ending link tag, </A>. When on an actual Web page, the hypertext link can be selected by a user, who will be sent to the place specified by the URL.

> Now for an exercise. Use your text editor and type in the basic page elements discussed earlier (<HTML> and </HTML>). Within the body of the page, type in a link tag pair using a real URL. Use one or both of the examples provided below. Once you have this page, save it with the name TEST.HTML and view it with your browser. (You do not have to be online for any of this.)
>
> **A Beginners Guide to HTML**
>   http://www.ncsa.uiuc.edu/General/Internet/WWW/HTMLPrimer.html
> **How to Publish On The Web**
>   http://www.thegiim.org/
> **The Bare Bones Guide to HTML**
>   http://werbach.com/barebones/
>
> See step #2 in the section below for help in finding and displaying the file. Once you've got this page up and running on your browser, try the hyperlink (online). Does it work? These URLs should take you to resources that will tell you more about Web publishing.

Home

So, you now have a homepage. Where do you put it to make it work? Well, this depends on whether you will have access to a Web server or not. The instructions that follow assume that you don't have access. Once you've completed your homepage, you'll have a file on either your hard drive or a floppy disk. The advantage of having your homepage on floppy disk is that you'll have a portable homepage that you can use on any computer. Here is a five-step procedure for finding and setting your default homepage to the one you've created.

1. After you finished the end-of-chapter exercise, you should copy your intended homepage to either a floppy disk or your hard drive.

2. With *Netscape Navigator*, you should be able to find a command under the File menu called *Open File...* . If you initiate this command, you'll get a dialog window where you can designate the file you wish to open. You want to open your homepage, which at the moment is not on a server. Select the HTML file for your homepage from either your floppy disk or hard drive and press the Return key. Because your file is written in HTML, your browser will open up and display it for you.

3. If it isn't set already, change your *Options* setting so that you can view the URLs of the pages that display on your browser. When this is set properly, you'll notice that your file's URL is listed something like this: file:/// ..../file.html. The protocol is no longer HTTP and there are more than two slashes.

4. Record this URL on a piece of paper so that you can refer to it in the next step. Every little slash and colon counts, so copy carefully in order to avoid a SNAFU (Situation Normal, All Fouled Up).

5. Earlier in this chapter, we discussed how to change the homepage designation for your browser. Go back to this section and use the description to open the *General Preferences* section of the *Options* menu. Select the *General* tab and enter the URL that you wrote down from step 4 into the homepage entry window.

Now that you've performed these steps, your browser should automatically jump you onto your homepage when you select the *Home* button from the toolbar. Give it a try. The procedures that you've just performed have allowed your browser to memorize the location of your homepage file. If you move it or erase it, then you'll have to do this all over again. Builds character—right?

Run Wild

If you want to learn more about tags, it's not hard to do so. Various places on the Web offer online tutorials on HTML design. Links to some of these places are found in Appendix II. As you learn more about the Internet and HTML, you'll discover ways to manually add pictures to your pages. Later, you'll even begin adding sounds and maybe a video clip or two.

We are at the beginning of a new way of learning and communicating. The tools available to you at this moment barely hint at what will be available to you a year from now. As you continue to journey through this new world, watch, look, and listen for new developments.

Section 4.2
Opening your own shop

A few of you may want to assemble and maintain an active Web server. If so, you'll need an extra measure of determination. A Web server is not necessarily easy to manage; you'll need to develop skills beyond those required to cruise the Net. However, doing so is well within your ability if you know how to get help and find resources. Also keep in mind that this enterprise will require a considerable amount of time, equipment, and money. But it will give you more control.

In order to get your project off the ground, you need to determine why you want a Web server. Being able to answer this fundamental question will help you sell your project to others, a key first step.

Now it is cheaper and easier to set up shop on the Net than it has been in years past. Computers are cheaper than ever and Web server software is nearly free; if you are in an academic setting, the software is free. But it still isn't cheap enough for you to do everything on your own. Get a sponsor. (This is where you need to sell your project idea.) Your school, your department, and probably some of your professors may be interested in setting up their own shop on the Net. They may have access to more financial resources than you do. Furthermore, they may lack the time and energy required to make a Web server work. You may want to suggest a partnership if you find an interested party. You will offer them your creative talent and time; in return, they will come up with the machine and the connection.

The shopping list
To set up a server on the Web, you need four basic things:

>   Computer
>   Software
>   Internet connection
>   A body

It's the last part of the list that stops most people. If you are going to setup a server, then you need someone who is willing and able to spend a considerable amount of time monitoring and maintaining it. You will probably be the one to do this work, so make sure you have the time and energy for it.

Unlike your desktop machine, a computer that is a server should be dedicated. Its sole function is that of being a server. Why? Because you may want to put important data on it that you don't want just anybody looking at—otherwise, data could be inadvertently erased. The following list contains suggestions for a basic server. Check with others to find out their opinions about server requirements; as always, consider your options and choose accordingly.

> Hardware:
> > Pentium computer (100 MHz is fine)
> > 36 MB RAM (minimum but 64 if you can get it)
> > 1 GB Hard Drive (start small and you can always add more)
> > CD-ROM Drive (easier for software installation)
> > Network interface card (this depends on your network)
> > Keyboard & monitor (this won't be a workstations, so these can be basic)
>
> Software:
> > Server operating system (Windows NT 4.0 Server)
> > Web server software (Microsoft Internet Information Server)
> > Mail and News server software (optional)
>
> Connection:
> > ISDN or T1 (24 hours-7 days a week)

I suggest a *Pentium* because they are relatively cheap and common. You don't need a really fast system like a *Pentium Pro* because a server's operations typically don't need speed; instead, they need gobs of bandwidth (Internet connection). Macintosh computers are also capable of acting as servers. However, if you want to be able to serve files to both Macintosh computers and PCs, then a PC is your best bet. Because many classrooms contain both types of computers, it is likely that you will need a PC server.

System memory is important if you want to run multiple elements like a mail server and a news server. The same goes for the hard drive. You'll need some storage, but keep in mind that you can always add to your system later on: Hard drives continue to get cheaper, faster, and larger. If you don't need extra memory now, then wait and save your money. The CD-ROM is probably essential if you have to do any software installation. Without it, you'll quickly learn how time consuming it can be to install software for one program from 50 floppy disks. The network interface card is used to connect your computer to your network (campus or outside provider). Take your provider's suggestions when purchasing this. It will save you a great deal of pain. The keyboard and monitor should be basic because you won't be doing any fancy graphic work. Save your money and buy a big monitor for your workstation.

Now for the software. I suggest Windows NT 4.0 Server as the base operating system: It is cheap, powerful, easy to use, growing in popularity (this means you'll be able to get lots of help), and can handle a mixed platform shop (a lab with both Macintosh computers and PCs). A great deal of software is currently being developed for this program, so you won't have a difficult time finding compatible software.

If you are planning to have a Web server, then you will need Web server software. Fortunately, *Microsoft's Internet Information Server* (IIS) is such software, and it is included with the WinNT system. Its advantages include price (it's free), easy set up, and ease of use. Netscape also provides Web server software for the WinNT platform. It, too, is free. I find the IIS software to be slightly more intuitive to use, but you may find otherwise. Try them both. If you need to set up a mail or news server, you will need additional software. These come in separate packages and are also commonly available from both Microsoft and Netscape. My advice is to start with a Web server and add more when you're comfortable with what you've created.

The components of a Web server are not unduly expensive—the computer and software will probably cost you less than $5000 in today's market. However, hardware and software are not your only expenses. A Web server without a network connection is nearly useless. Even if you want to setup an internal network (Intranet), you'll need the basic wires and connections to other computers. Keep in mind that your school may be able to make a Web server possible by providing computer resources that would otherwise be too expensive. If your school cannot offer such assistance, you will have to rent Internet access on a monthly basis. This will probably cost a considerable amount of money. Thus, be sure that you have a compelling need for a server before you begin to assemble one.

If you do put together a server, you may need additional equipment to ensure its security: for example, a tape backup system and an un-interruptible power supply (UPS). If your stuff is really important, you may want to even consider locking everything in a back room with a redundant server. Of course, this is only if you have valuable information.

If you're determined to set up a Web server, then you need these URLs. They'll take you to sites on the Web where you'll learn about setting up, maintaining, and administering a Windows NT server. You'll also be able to download freeware versions of Microsoft's, Netscape's, and other companies' Web server software.

Web66 Windows NT Cookbook
   http://web66.coled.umn.edu/WinNT/CookBook/Default.html
Windows NT Utilities
   http://www.ime.net/~peteb/
Netscape's Education Program
   http://home.netscape.com/comprod/server_central/edu_drive.html
Microsoft's Education Program
   http://www.microsoft.com/education/hed/default.htm

We began this journey by exploring the great Information Superhighway, the Internet, and now you've just learned a little bit about how to jump from the driver seat into the Web mechanic's garage. The transition isn't easy but it is rewarding. By now, you've seen many different resources on the Internet. Some of them are very well designed, with valuable, accurate information; some may seem like a waste of bandwidth. As you continue along this road, consider what positive things you can contribute to this revolution. I hope that you now find the Internet more understandable, and that you'll go on to learn more from other resources, including books, online sites, and your fellow travelers. As always, be aware of your options, consider your choices, and make an informed decisions.

Activity: Make yourself at home!

If you've been working as you read this chapter, then you probably have something that is pretty close to a home page for yourself. Congratulations! Have you considered adding your important bookmarks to your homepage? As a teacher and administrator of a student network, I found that students' travels often led to the gradual and insidious buildup of cluttersome bookmarks. As a student, you may be tempted to change the browser in your classroom to meet your personal needs. However, this can create extra work for your intructors, who never know what to expect when they teach a class or use a browser. Likewise, some of your fellow students may find your changes to be confusing or bothersome. Your homepage may be the solution.

With the information provided in Appendix II, you will be able to build a student homepage to be kept on a floppy disk, if you haven't done so already. Keep in mind that you can use your homepage as a portable resource disk on most networked computers or you can post it with your ISP and have world wide access. The only disadvantage to keeping your homepage on a floppy disk is that you can't jump between Macintosh and Windows operating systems. Perhaps software developers will be able to overcome this difficulty in the future.

In Appendix II, I used several different fonts in order to distinguish various elements of the homepage template. *Large italic font* is used to represent the text that you are to replace with your information. HTML tags are represented in a <SMALLER FONT, ALL IN UPPER-CASE AND WITH ANGLE BRACKETS>. Text that is not an HTML tag and does not need to be changed is in a regular size and font.

Now, take the time to really dig into your homepage. You can either use yours or the one I've provided in the appendix. You can use it to post it on your campus Web server, your ISP, put it on your home computer, or place it on a floppy for portable use. Give the file a name with the suffix html (HOMENAME.HTML). Just remember to show some style and personality!

You have my e-mail address. Please drop me a note.

Appendix I
It's news to me
Mathematics Newsgroups

Think of the list that follows as a starter kit for newsgroups. However, keep in mind that this is not a complete list; also, some of the newsgroups listed will undoubtedly disappear in the near future. Furthermore, many new groups will be born in this same time period. One of them might even be yours. As I've said before, it's a changing world.

The names generally describe the topic of discussion. You will find that some groups are very interesting but others are not. Subscribe to several of the groups and eavesdrop on their conversations for a while before you jump in. This will help you discover the general attitude of the group. I also suggest that you read the FAQ and any postings directed to new participants.

sci.answers	sci.math
sci.math.symbolic	sci.math.num-analysis
sci.math.research	alt.math.undergrad
sci.fractals	sci.nonlinear
comp.theory.cell-automata	comp.theory.dynamic-sys
sci.logic	alt.algebra.help
geometry.announcement	geometry.college
geometry.forum	geometry.institutes
geometry.puzzles	geometry.research
geometry.software.dynamic	comp.ai.alife
comp.theory.self-org-sys	comp.soft-sys.math.mathematica
sci.physics.computational.fluid-dynamics	sci.op-research
sci.stat.math	sci.research
sci.research.careers	sci.crypt
sci.misc	misc.education.science

Appendix II
Stepping Out

Student Homepage Template:

```
<HTML>
<HEAD>
    <TITLE>Your Homepage</TITLE>
</HEAD>
<BODY>
<CENTER>
    <H1>Your Name</H1>
    <H2>Your Title, Major, or Philosophy</H2>
    <H3>
    <ADDRESS>
        Your address<BR>
        May<BR>
        Go<BR>
        Here<BR>
        <P>
        Your e-mail address<BR>
    </ADDRESS>
    </H3>
</CENTER>
<HR>

<DL>
    <H2>This Is Your Life:</H2>
    <DD>Just say something about yourself. After you've added something to your homepage file, compare it to what actually shows on your browser. You'll notice that most of the HTML code is not visible.
    <P>
    <DD>You can have as many paragraphs as you wish. Here's another. Enjoy.
    <P>
</DL>
```

```
<DL>
    <H2>Important Resources</H2>
    <DD>
        <A HREF="http://www.ssnewslink.com">
            NewsLink from Simon & Schuster
        </A>
</DL>
<HR>
<DL>
    <H2>Handbooks and Manuals</H2>
    <DD>
        <A HREF="http://www.matisse.net/files/glossary.html">
            Glossary of Internet Terms
        </A>
    <DD>
        <A HREF="http://home.netscape.com/eng/mozilla/3.0/handbook/">
            Netscape Handbook
        </A>
    <DD>
        <A HREF="http://home.netscape.com/eng/mozilla/3.0/handbook/authoring/navgold.htm">
            Netscape Gold Authoring Guide
        </A>
</DL>
<HR>
<DL>
    <H2>Directories and Search Engines: </H2>
    <DD>
        <A HREF="http://www.lycos.com">
            Lycos Search Engine
        </A>
    <DD>
        <A HREF="http://www.Webcrawler.com/">
            WebCrawler Search Engine
        </A>
    <DD>
        <A HREF="http://www.yahoo.com/">
            Yahoo Directory
        </A>
</DL>
<HR>
<DL>
    <H2>Design and Publish: </H2>
    <DD>
        <A HREF="http://www.ncsa.uiuc.edu/General/Internet/WWW/HTMLPrimer.html">
            A Beginner's Guide to HTML
        </A>
    <DD>
        <A HREF="http://werbach.com/barebones/">
            The BareBones Guide to HTML
        </A>
    <DD>
        <A HREF="http://www.thegiim.org/">
            How to Publish On The Web
        </A>
```

```
            <DD>
                <A HREF="http://home.netscape.com/assist/net_sites/starter/samples/index.html">
                        Netscape's Gold Rush Tool Chest
                </A>
</DL>
<HR>
<H3>
        Your Name <A HREF="mailto:Name@server.edu">e-mail address</A>
</H3>

</BODY>
</HTML>
```

Appendix III
Useful URLs
Mathematics Web Sites

History of Mathematics Home Page
http://aleph0.clarku.edu/~djoyce/mathhist/mathhist.html

Biographies of Women Mathematicians
http://www.scottlan.edu/lriddle/women/women.html

Mathematics Department Web Servers
http://www.math.psu.edu/OtherMath.html

Mathematical Quotations Server
http://math.furman.edu/~mwoodard/mquot.html

WHO's On-line
http://www.math.psu.edu/WHO/math.html

E-Math
http://www.ams.org/

Internet Center for Mathematics Problems
http://www.mathpro.com/math/mathCenter.html

Math Forum
http://forum.swarthmore.edu/

Knot A Braid of Links
http://camel.math.ca/Recreation/kabol/knotlinks.html

JAVA Gallery of Interactive On-Line Geometry
http://www.geom.umn.edu/java/

The Prime Page
http://www.utm.edu/research/primes/

Virtual Reality Polyhedra
http://www.li.net/~george/virtual-polyhedra/vp.html

Mathematical Games, Toys, and Puzzles
http://www.cs.duke.edu/~jeffe/mathgames.html

Primordial Soup Kitchen
http://math.wisc.edu/~griffeat/welcome.html

Mathematics Archives
http://archives.math.utk.edu/

Glossary
It's all Greek to me

ActiveX

This is a resource developed by *Microsoft* to extend the function of their *Internet Explorer* software.

Archie

This is a search tool used to find resources that are stored on Internet-based FTP servers. Archie is short for Archive because it performs an archive search for resources. (See *FTP* and *Server*.)

AVI

This stands for Audio/Video Interleaved. It is a Microsoft Corporation format for encoding video and audio for digital transmission.

Background

This refers to an image or color that is present in the background of a viewed Web document. Complex images are becoming very popular as backgrounds but require a great deal more time to download. The color of default background can be set for most Web browsers.

Bookmark

This refers to a list of URLs saved within a browser. The user can edit and modify the bookmark list to add and delete URLs as the user's interests change. Bookmark is a term used by Netscape to refer to the user's list of URLs; *Hotlist* is used by Mosaic for the same purpose. (See *Hotlist, Mosaic,* and *URL*.)

Browser

This is a software program that is used to view and browse information on the Internet. Browsers are also referred to as clients. (See *Client*.)

Bulletin Board Service

This is an electronic bulletin board. It is sometimes referred to as a BBS. Information on a BBS is posted to a computer where people can access, read, and comment on it. A BBS may or may not be connected to the Internet. Some are accessible by modem dial-in only.

Cache
This refers to a section of memory that is set aside to store information that is commonly used by the computer or by an active piece of software. Most browsers will create a cache for commonly accessed images. An example might be the images that are common to the user's homepage. Retrieving images from the cache is much quicker than downloading the images from the original source each time they are required.

Chat room
This is a site that allows for real-time person-to-person interactions.

Clickable image (Clickable map)
This refers to an interface used in Web documents that allow the user to click, or select, different areas of an image and receive different responses. Clickable images are becoming a popular way to offer a user many different selections within a common visual format.

Client
This is a software program used to view information from remote computers. Clients function in a Client-Server information exchange model. This term may also be loosely applied to the computer that is used to request information from the server. (See *Server*.)

Compressed file
This refers to a file or document that has been compacted to save memory space so that it can be easily and quickly transferred through the Internet.

Download
This is the process of transferring a file, document, or program from a remote computer to a local computer. (See *Upload*.)

E-mail
This is the short name for electronic mail. E-mail is sent electronically from one person to another. Some companies have e-mail systems that are not part of the Internet. E-mail can be sent to one person or to many different people. (I sometimes refer to this as JunkE-mail.)

FAQ
This stands for Frequently Asked Questions. A FAQ is a file or document where a moderator or administrator will post commonly asked questions and their answers. Although it is very easy to communicate across the Internet, if you have a question, you should check for the answer in a FAQ first.

Forms

This refers to an interface element used within Web documents to allow the user to send information back to a Web server. With a forms interface, the user is requested to type responses within entry windows to be returned to the server for processing. Forms rely on a server computer to process the submittals. They are becoming more common as browser and server software improve.

FTP

This stands for File Transfer Protocol. It is a procedure used to transfer large files and programs from one computer to another. Access to the computer to transfer files may or may not require a password. Some FTP servers are set up to allow public access by anonymous log-on. This process is referred to as Anonymous FTP.

GIF

This stands for Graphics Interchange Format. It is a format created by CompuServe to allow electronic transfer of digital images. GIF files are a commonly-used format and can be viewed by both Mac and Windows users.

Gopher

Is a format structure and resource for providing information on the Internet. It was created at the University of Minnesota.

GUI

Is an acronym for Graphical User Interface. It is a combination of the appearance and the method of interacting with a computer. A GUI requires the use of a mouse to select commands on an icon-based monitor screen. Macintosh and Windows operating systems are examples of typical GUIs.

Helper

This is software that is used to help a browser view information formats that it could not normally view.

Homepage

In its specific sense, this refers to a Web document that a browser loads as its central navigational point to browse the Internet. It may also be used to refer to as Web page describing an individual. In the most general sense, it is used to refer to any Web document.

Hotlist

This is a list of URLs saved within the Mosaic Web browser. This same list is referred to as a *Bookmark* within the Netscape Web browser.

HTML
An abbreviation for HyperText Markup Language, the common language used to write documents that appear on the World Wide Web.

HTTP
An abbreviation for HyperText Transport Protocol, the common protocol used to communicate between World Wide Web servers.

Hypertext
This refers to text elements within a document that have an embedded connection to another item. Web documents use hypertext links to access documents, images, sounds, and video files from the Internet. The term hyperlink is a general term that applies to elements on Web pages other than text.

Inline image
This refers to images that are viewed along with text on Web documents. All inline images are in the GIF format. JPEG format is the other common image format for Web documents; an external viewer is typically required to view such documents.

Java
This is an object-oriented programming language developed by Sun Microsystems.

JavaScript
This is a scripting language developed by Netscape in cooperation with Sun Microsystems to add functionality to the basic Web page. It is not as powerful as Java and works primarily from the client side.

JPEG
This stands for Joint Photographic Experts Group. It is also commonly used to refer to a format used to transfer digital images.

Jughead
This is a service for performing searches on the Internet. (See *Archie* and *Veronica*.)

Mosaic
This is the name of the browser that was created at the National Center for Supercomputing Applications. It was the first Web browser to have a consistent interface for the Macintosh, Windows, and UNIX environments. The success of this browser is responsible for the expansion of the Web.

MPEG

This stands for Motion Picture Experts Group. It is also a format used to make, view, and transfer both digital audio and digital video files.

Newsgroup

This is the name for the discussion groups that can be on the Usenet. Not all newsgroups are accessible through the Internet. Some are accessible only through a modem connection. (See *Usenet*.)

Plug-in

This is a resource that is added to the Netscape to extend its basic function.

QuickTime

This is a format used by Apple Computer to make, view, edit, and send digital audio and video.

Server

This is a software program used to provide, or serve, information to remote computers. Servers function in a Client-Server information exchange model. This term may also be loosely applied to the computer that is used to serve the information. (See *Client*.)

Table

This refers to a specific formatting element found in HTML pages. Tables are used on HTML documents to visually organize information.

Telnet

This is the process of remotely connecting and using a computer at a distant location.

Upload

This is the process of moving or transferring a document, file, or program from one computer to another computer.

URL

This is an abbreviation for Universal Resource Locator. In its basic sense it is an address used by people on the Internet to locate documents. URLs have a common format that describes the protocol for information transfer, the host computer address, the path to the desired file, and the name of the file requested.

Usenet
This is a world-wide system of discussion groups, also called newsgroups. There are many thousands of newsgroups, but only a percentage of these are accessible from the Internet.

Veronica
Believe it or not, this is an acronym. It stands for Very Easy Rodent Oriented Net-wide Index to Computerized Archives. This is a database of menu names from a large number of Gopher servers. It is a quick and easy way to search Gopher resources for information by keyword. It was developed at the University of Nevada.

VRML
This stands for Virtual Reality Markup Language. It was developed to allow the creation of virtual reality worlds. Your browser may need a specific plug-in to view VRML pages.

WAIS
This stands for Wide Area Information Servers. This is a software package that allows the searching of large indexes of information from the Internet.

- ## WAV
  This stands for Waveform sound format. It is a Microsoft Corporation format for encoding sound files.

Web (WWW)
This stands for the World Wide Web. When loosely applied, this term refers to the Internet and all of its associated incarnations, including Gopher, FTP, HTTP, and others. More specifically, this term refers to a subset of the servers on the Internet that use HTTP to transfer hyperlinked documents in a page-like format.